Improving Global Worker Health and Safety Through Collaborative Capacity Building Initiatives

Improving Global Worker Health and Safety Through Collaborative Capacity Building Initiatives

Edited by
Thomas P. Fuller

CRC Press
Taylor & Francis Group
Boca Raton London New York

CRC Press is an imprint of the
Taylor & Francis Group, an **informa** business

First edition published 2022
by CRC Press
6000 Broken Sound Parkway NW, Suite 300, Boca Raton, FL 33487-2742

and by CRC Press
2 Park Square, Milton Park, Abingdon, Oxon, OX14 4RN

© 2022 Taylor & Francis Group, LLC

CRC Press is an imprint of Taylor & Francis Group, LLC

ISBN: 9780367459185 (hbk)
ISBN: 9781032034980 (pbk)
ISBN: 9781003026471 (ebk)

Typeset in Times
by codeMantra

Contents

Preface

Occupational hygiene (OH) is the profession of anticipating, recognizing, evaluating and controlling workplace risks. Although there are over 8,000 members of the American Industrial Hygiene Association (AIHA), the demand for occupational hygienists far outpaces the US educational system's abilities to provide qualified graduates in the field. In the United States, this is partly a marketing problem. Almost every fifth grader knows the title "nurse" and generally knows what nurses do. But unfortunately, not everyone is as familiar with what an occupational hygienist does. Largely as a result, even though many OH university programs exist and graduates from them find rewarding jobs upon graduation, universities cannot attract enough students into the field of study to fill the job market demand.

In many economically developing countries (EDCs) and in nations without robust OH educational programs, the shortage of qualified occupational hygienists is even more alarming. As a result, employees in these countries suffer higher rates of workplace injury, illness, and fatality. Damage to valuable infrastructure, the environment, and public health also occurs in areas without adequate professional OH support.

In understanding the importance of occupational hygienists and the role they play in worker health and well-being, many nonprofit organizations have been created to directly address some of the more egregious problems in EDCs. Through an elaborate, but transparent, network of collaboration and communication, numerous educational, research, and development projects have been completed to assist EDCs in the expansion of OH capacity over the past several years. The stories of how these groups formed, and the work they continue to do, are of interest to those currently practicing OH and are valuable tools for students entering the field, as well as experienced and emeritus professionals, who also want to do philanthropic international work.

This book outlines a broad variety of capacity-building projects in OH. It also introduces some detailed experiences of occupational hygienists working abroad, or working on special global topics including a discussion of how global trade agreements influence the practice and policy of OH, the special needs of informal workers who do not have "typical" employers, and special challenges of creating international experiences for university students as a means to understanding the special occupational health and safety challenges in other countries. This is also a form of building competency and increasing capacity in global OH.

This book will be of most interest and use to OH professionals with several years of experience who are interested in extending and broadening their careers with philanthropic work abroad. Professionals who have spent their lives protecting workers from occupational hazards know the value of their profession and are looking for a way to share their expertise and knowledge. The need lies in knowing which nonprofit organizations are doing what type of work, to best match the professional's interest with the needs of the organizations. Knowledge exchange through networking and collaboration between these organizations and other professional and governmental organizations is an excellent way to begin to address the global need for occupational hygienists. This book can serve as a link between these organizations and a one-stop source of related information.

The concepts for the chapters in this book are, in many cases, a review of the creation and operation of the organizations being discussed. Some aspects of those stories will be timeless, for example, the challenges associated with funding sources, or organizational governance. Other aspects of the chapters look toward current projects and future work. It seems that most planning, and thus relevance of the topics, might extend 5 years. Some aspects of some of the chapters may be timeless, such as the importance of fair-trade policies or robust regulatory systems and the impacts of OH.

Many of the authors and co-authors of the chapters in this book are members of the AIHA International Affairs Committee (IAC). Through regular monthly IAC meetings, many of the projects discussed in this book are presented and communicated to others. Collaborative projects are inspired through open discussions, and a form of synergy is created. In a way, this book began as a forum to showcase some of the projects and collaborations that flowered through regular support of the IAC.

Several authors are members of other international OH professional organizations, many of which (including the AIHA) are members of the International Occupational Hygiene Association (IOHA). The IOHA is another forum in which many of these organizations have communicated and collaborated on a regular basis over the past few decades.

The time frame for conception of the book, to receipt of all chapter drafts, was just over 1 year. Editing and publication was scheduled to take another 6 months. The numerous authors worked in parallel on their respective chapters. Both the introduction and conclusion chapters were written after all the chapter drafts were received and reviewed. Each chapter was peer-reviewed by the editor and two additional professionals with specialized experience and expertise in international OH issues and topics.

Acknowledgments

I would like to express my sincere appreciation to each of the co-authors who contributed to portions of this book. I would like to thank the individuals who provided chapter reviews and suggestions for improvements in various sections of the book: David Hicks, Stephen Chiusano, Michael Vangeel, David Zalk, Karen Gunderson, Milja Koponen, Nathalie Argentin, Kathleen Bradford, and Ruth Jimenez. I would also like to thank those who have made a significant impact on my understanding of global occupational health and safety issues over the past two decades: Marcos Domingos da Silva, Dave Marsh, Kul Garg, Lydia Renton, Rene Leblanc, Andrea Hiddinga, Peter Jacobs, and Sergio A. Caporali Filho. I also thank Larry Sloan and the staff at the American Industrial Hygiene Association for their continued support of international worker health and safety initiatives and collaborations.

Editor

Dr. Thomas P. Fuller is a Professor in the ABET Accredited Occupational Safety and Health Program at Illinois State University. He earned a Doctor of Science from the University of Massachusetts Lowell in Industrial Hygiene, a Master of Science in Public Health from the University of North Carolina, and a Master of Business Administration from Suffolk University. He is a Certified Industrial Hygienist (CIH), a Certified Safety Professional, and a Fellow of the American Industrial Hygiene Association (FAIHA). Dr. Fuller is the current President of the International Occupational Hygiene Association (IOHA) and Chair of the IOHA Education Committee. He is a past Chair of the AIHA International Affairs Committee (IAC) and is currently Chair of the IAC Emerging Economy Microgrants Subcommittee. Dr. Fuller is a Contributing Editor for the *American Journal of Nursing*, a member of the Editorial Advisory Board of the National Safety Council, and a member of the Editorial Advisory Panel of the journal of Occupational Hygiene Southern Africa. As a member of the International Commission on Occupational Health, he serves on the Industrial Hygiene Committee and the Working Group on Infectious Occupational Agents. He is also a Director on the Board of the Occupational Hygiene Training Association (OHTA).

Contributors

Roger Alesbury, a founder, director and former chair of the OHTA, Roger has more than 45 years' experience as an occupational hygienist, mostly working internationally in the oil, gas, and chemical industries. Prior to retirement in 2010, he was Head of Industrial Hygiene for BP, with international responsibilities including the development of global occupational hygiene policy, capability, and resources.

He has an MSc in occupational hygiene from the London School of Hygiene and Tropical Medicine and the British Occupational Hygiene Society (BOHS) Diploma of Professional Competence. Formerly a Chartered Occupational Hygienist, Chartered Fellow of the Faculty of Occupational Hygiene, and Chartered Fellow of the Institution of Occupational Safety and Health, with retirement his registrations have now lapsed. He is a past President of BOHS and past President of the Institute of Occupational Hygienists. He has served on numerous professional, industry, and government committees in the EU and the UK. In 2011, he was awarded the ACGIH William Steiger Memorial Award for contributing to advancements in occupational safety and health (OSH); the 2016 AIHA Yant Award for outstanding contributions to the industrial hygiene (IH) profession; and the 2017 IOHA Lifetime Achievement Award to mark continuing outstanding contribution to occupational hygiene.

Ulrike Bollmann is Head of International Cooperation at the Institute for Work and Health (IAG) of the German Social Accident Insurance (DGUV). After studying education and philosophy, she conducted research in the field of schools and further education and developed curricula for various professions. In 1999, she began her work in the IAG of the DGUV. From 2002 to 2004, she was a national seconded expert at the European Agency for Safety and Health at Work (EU-OSHA) in Spain. Since 2005, Ulrike has been the founder and coordinator of the European Network Education and Training in Occupational Safety and Health (ENETOSH). She organized several major international conferences in cooperation with the World Health Organization (WHO), International Labour Organization (ILO), International Social Security Association (ISSA), EU-OSHA, and International Association of Labour Inspection (IALI). She was responsible for an empirical study on mainstreaming OSH in education, conducted a joint research project with Korea Occupational Safety and Health Agency (KOSHA) to develop leading indicators. She recently published a book on OSH competences together with George Boustras. Ulrike is a member of the Editorial Board of *Safety Science*, a member of the Scientific Committee on Education and Competency Development of OSHAfrica, and a member of the Steering Group of the EU-OSHA campaign on MSDs 2020–2022.

Garrett D. Brown is a CIH and earned a Master's in Public Health from the University of California at Berkeley. Brown worked as a field compliance officer for the California Division of Occupational Safety and Health (Cal/OSHA) for 18 years and then worked for three years as the Special Assistant to the Chief of the Division before retiring in 2014. Brown founded the Maquiladora Health and Safety Support Network made up of 300 occupational health professionals in 1993 and has been the volunteer coordinator since then. Brown has organized projects to increase the OHS knowledge and capacity of grassroots worker and community organizations in Mexico, Central America, Indonesia, China, and Bangladesh. Brown has written extensively in professional journals and trade publications about workplace health and safety in global supply chains. Since 2018, Brown has taught classes on global OHS issues at UC Berkeley's School of Public Health.

Andrew Cutz is a CIH and a Fellow of the AIHA with more than 25 years of progressive experience. He completed his Bachelor of Science degree in Health Sciences at the University of Toronto with subsequent postgraduate medical Diploma in Industrial Health (DIH) from the University of Toronto's Occupational and Environmental Health Unit. Cutz worked as an Environmental Project Coordinator at the University of Toronto and then joined the Canadian Centre for Occupational Health and Safety (CCOHS) in Hamilton, Ontario, where he was Project Officer in the Physical Agents Group for 5 years, and later the Ontario Ministry of Labour as their Occupational Hygiene Consultant for North Eastern Ontario for 7 years. Most recently, he was an instructor within the Graduate Environmental and Occupational Health Track at the St. George's Medical School in Grenada, West Indies. Cutz acted as a team leader on the EU-funded Technical Assistance for Development of Regional Laboratories Project for the Occupational Health and Safety Centre (İSGÜM) in Ankara, Turkey (2010–2012). He also operated a private IH consulting business for the past 10 years and worked for environmental consulting companies, providing comprehensive and timely IH services, and occupational health and safety training to institutional and corporate clients in the Greater Toronto Area and across the Ontario's Golden Horseshoe.

Barbara J. Dawson is a past President of the AIHA and a past Chair of the American Board of Industrial Hygiene (ABIH). She is a CIH, a Certified Safety Professional, and a Fellow of the AIHA. She is the Global Occupational Hygiene Competency Leader and an Environmental, Health and Safety Fellow for the DuPont Company where she has worked for more than 30 years. She was the recipient of the 2020 American Chemistry Council Member Company Responsible Care Employee of the Year award. Barbara earned an MS degree in Environmental Health (Industrial Hygiene) from Temple University in Philadelphia, Pennsylvania, and a BS degree in biology from Muhlenberg College in Allentown, Pennsylvania.

Claus Dethleff is a lecturer in media design, web designer, editor, and architect. He earned a diploma in architecture and has been working in the media sector for more than 20 years now. Currently, he is working as web editor, architect, and lecturer in Dresden, Germany. Dethleff has worked in the field of OSH as the chief editor of the web platform of the ENETOSH (www.enetosh.net) since 2015.

Deborah F. Dietrich earned a Master of Science in Industrial Hygiene from the University of Texas School of Public Health. She is a CIH. She is an Honorary Fellow of the Australian Institute of Occupational Hygienists (AIOH) and has received the award for Promotion of Occupational Hygiene in Latin America by the Association of Brazilian Occupational Hygienists. She is a past Director of the AIHA. Deborah worked as Senior Vice President and Corporate Industrial Hygienist of SKC, Inc., from 1984 to 2020. In this role, she prepared technical publications and training tools on all aspects of sampling for use by hygienists globally through the SKC website and SKC worldwide distribution network. She also conducted training on air sampling and provided technical support to clients around the world.

Dr. Ilise L. Feitshans, an international lawyer and former international civil servant at the United Nations, Geneva, Switzerland, earned her Master's of Science in Public Health from Johns Hopkins University, Baltimore, Maryland, USA, and Doctorate in International Relations from Geneva School of Diplomacy, Switzerland. Her books *Walking Backwards to Undo Prejudice: Report of the US Capitol Conference Including Disabled Students: What Works, What Doesn't* (Lambert Academic Publishers 2019) and *Global Health Impacts of Nanotechnology Law* (Panstanford 2018) address disability rights and new assistive technologies. She is a member of the US Supreme Court bar and was acting director of the Legislative Drafting Research Fund, Columbia University School of Law, New York, USA. Ilise served as coordinator for the *ILO Encyclopedia of Occupational Health and Safety* in Geneva, Switzerland. A graduate-cum-laude of Barnard College of Columbia University, New York, USA, she was also visiting scientist at the Institute for Work and Health, University of Lausanne, Switzerland (2011–2014). She was honored among "100 Women Making a Difference in Safety, Health and Environment Professions" by the American Society of Safety Engineers in 2011 and received the Ms-JD.org Superwomen Award in 2016.

Ilda Luísa Figueiredo earned a degree in Psychology and a postgraduate in Psychology and in Didactic Perspectives in Curricular Areas. She is a senior technician of the Directorate-General for Education working in Citizenship Education, namely in the domains Labour World and Entrepreneurship Education, among others, and she is a co-author of guidelines and other publications in this field. In this context, she has been representing the Directorate General of Education – Ministry of Education in various working groups at national and international levels.

Dr. Paul Leonard Gallina is a Professor in the Williams School of Business, Bishop's University, Sherbrooke, Quebec. There he teaches courses in employment law, labor relations, and occupational health and safety. His research interests include employment standards, comparative regulatory approaches to occupational health and safety, and occupational disability management. He has acted as a consultant to the Government of Canada, the European Union, and various public and private corporations.

Richard Hirsh is certified in the comprehensive practice of IH and currently works as Sr. Director, EH&S, for Nektar Therapeutics, a biopharmaceutical firm with operations in the United States and India. Before that, Richard served as Global Safety and IH Manager for Stiefel Laboratories and previously worked for Rohm and Haas Company in the Global EHS Department for 20 years. Richard is an AIHA Fellow and serves on the board of the California Industrial Hygiene Council (CIHC). He served on the AIHA International Scientific Planning Committee for the IOHA Conference in 2018. He is a past chair of the IAC of the AIHA and served as chairperson for the planning committees of the AIHA Asia Pacific Conferences held in Malaysia and Singapore. Richard is a past President of the Northern California Section of the AIHA and serves as founder and chair of the Developing World Outreach Initiative. He also served on the Board of Directors for the US branch of Workplace Health Without Borders (WHWB) and continues to serve on the Advisory Committees for the Centers for Occupational and Environmental Health at both the University of California and the University of Michigan. Richard attended UC Berkeley, School of Public Health, Environmental Health Sciences Program where he earned his M.P.H. in 1986. He earned a biology degree from Lafayette College.

Chris Laszcz-Davis has more than 35 years of executive management, professional, technical, operational, and management consulting experience in environmental affairs, occupational health and safety, operational integrity, risk management, and product stewardship in industry and government. She is currently President of The Environmental Quality Organization LLC (EQO). As former Corporate Vice-President of Environmental Affairs, Health, Safety & Operational Integrity for Kaiser Aluminum & Chemical Corporation, Chris helped lead the company initiatives to reshape its EH&S culture, expand its global EH&S horizons to address the full life cycle of its organization's aluminum and chemical operations, and integrate its EH&S workings into the fabric of the line organization. Chris is an AIHA Fellow, a recipient of AIHA's Kusnetz and Alice Hamilton Awards, and a recipient of the Yuma Pacific Southwest Section's Clayton Award and Northern CA Section's Dr. Christine Einert Award. Chris has served as a board member of AIHA, ABIH, ACGIH, and American Industrial Hygiene Foundation (AIHF); President of American Academy of Industrial Hygiene; US Department of Commerce Appointee of Board of Examiners for Malcolm Baldrige National Quality Award; President of both Northern California and Yuma Pacific Southwest local sections; President of CIHC; and ABIH Appointee of IOHA's National Accreditation Recognition Committee. Chris was until recently a member of NIOSH's Board of Scientific Counselors (BSC); is a member of Cal/OSHA's Standards Board; is Co-Chair of the global OHTA; is a member of ASSP's Governmental Affairs Committee (GAC); and is recent Chair of the ASSP Task Force on Total Worker Health (TWH). Chris lectures frequently and authors numerous articles (recent ILO presentation and article on "The New World Battleground with COVID-19: Challenges, Partnerships, Impact and Business") and signature chapters (with recent PATTY chapters on Human Health Risk Assessment and Emergency Preparedness, Response & Recovery).

Marianne Levitsky was founding President of WHWB, a nonprofit organization that engages volunteers in promoting occupational health for workers everywhere. A senior associate with ECOH Management, a Canadian consulting firm, she was previously Director of the Prevention Best Practices for the Ontario Workplace Safety and Insurance Board, an occupational hygienist with the Ontario Ministry of Labour, and a co-founder of the Toronto Workers' Health and Safety Legal Clinic. She is adjunct faculty at the University of Toronto and was a member of the Toronto Board of Health. She has served as chair of the AIHA IAC and is an AIHA Fellow. She is a recipient of the Hugh Nelson Award from the Occupational Hygiene Association of Ontario and the Yant Award from the AIHA.

Dr. Zack Mansdorf is a consultant in EHS and sustainability. He is a former Managing Principal Consultant with BSI EHS Services and Solutions. Prior to BSI, Zack was Senior Vice President-Safety, Health, and Environment for L'Oreal Worldwide. He has also held several other senior level positions in the US Army (retired Lt. Col.) Goodyear, Midwest Research Institute, Clayton Environmental Consultants, Liberty International Risk Services, and Arthur D. Little Company. Zack earned a PhD in Environmental Engineering (University of Kansas) with Master's degrees in EHS and Safety from the University of Michigan and Central Missouri State University. He is certified in all three areas (QEP, CSP, and CIH). He is a past President of the Academy of Industrial Hygiene and the AIHA. He is a past board member of the Board of Certified Safety Professionals (BCSP) and past President of the AIHF. He is a Fellow of ASSP, AIHA, ASTM, and the Controlled Release Society. Zack also served on the ASSP Council on Professional Affairs (COPA) and currently serves on the Government Affairs Committee and the Z16 and TC283 committees. He is a past Director of the Center for Safety and Health Sustainability. He is a current Director in the OHTA.

Nancy M. McClellan, prior to returning to independent consulting practice, McClellan practiced IH management on a global scale for both AbbVie, a major producer of highly potent biopharmaceuticals, and LafargeHolcim, the world's largest building materials corporation. She has invested more than 25 years in practicing occupational hygiene in a wide variety of high-hazard industries. As a self-employed consultant for more than 15 years, she provides legal expert, management, training, and select field services to sectors such as the legal, pharmaceutical, automotive, military, hospital, electronic, food, chemical, and housing industries. In her volunteer professional life, McClellan currently serves as the Chair of the University of Michigan Graduate School of Public Health External Advisory Board to help lead and shape the academic programs for today's industry demands. She recently served on the Board of Directors for both the AIHA and the ABIH to address global IH competency by forming and leading the ABIH Futures Committee. As the past Co-Chair and leader for the globally recognized OHTA for 13 years, she facilitated the growth of the organization to train over 10,000 IH students worldwide and ensure the program continues growing with a sustainable system of governance. She has been an active member of the British and Irish Occupational Hygiene Societies and a number of AIHA Chapters as a highly rated AIHA Distinguished Lecturer and organizational liaison to AIHA and OHTA.

Dr. Dingani Moyo is an OSH consultant, occupational medicine specialist, and lecturer at Midlands State University in Zimbabwe and the University of Witwatersrand in South Africa. He has more than 20 years of experience in the practice of occupational health and safety in the Southern African Development Community (SADC). He earned a Master's degree in Occupational Health and Safety, Master's degree in Health Services Management (Newcastle, Australia), and higher specialist qualifications in occupational medicine – Fellowship in Occupational Medicine from the Royal College of Physicians in Ireland. He also earned a Bachelor of Medicine and Surgery degree (Zimbabwe).

Tuan N. Nguyen has more than 30 years of experience as an IH consultant providing safety and health services to businesses in high-hazard industries for the largest workers' compensation insurance company in California. Tuan had worked at the Air Pollution Health Effects Laboratory at the University of California Irvine Medical Center where he helped coordinate and conduct research in air pollution inhalation toxicology. He served as a team leader for the Nail Salon Health Hazards Evaluation Project at the State Compensation Insurance Fund, had co-authored several scientific publications in air pollution and nail salon research studies, and has made presentations at conferences in the United States and Vietnam. Tuan had organized and co-taught the OHTA courses at the National Institute of Occupational and Environmental Health in 2016 and 2018 in Hanoi and Ho Chi Minh City, Vietnam. He is currently a roster candidate of the Fulbright Specialist Program. He currently serves as the AIHA Ambassador to Vietnam and as the Orange County AIHA Chapter Secretary. In 2018, Tuan was awarded with the AIHA's Fellow designation. In 2016, he co-founded the Vietnamese Industrial Hygiene Association (VIHA), and currently serves as a member of the board of directors. He is a board member of the IOHA and Vice President of WHWB-US.

Dr. Mary O'Reilly earned her Doctorate in Human Anatomy and Cell Biology from the University of Michigan and is a CIH and a certified professional ergonomist. She performed quantitative human risk assessments for the US EPA as an environmental toxicologist at Syracuse Research Corporation and worked as an industrial hygienist and as an environmental specialist for New York State for many years. She is adjunct faculty at the University at Albany School of Public Health, a member of the Institute for Health and the Environment and president of ARLS Consultants. Mary has been a member of the AIHA for more than 30 years and part of the leadership team for several AIHA committees. She also chaired the Occupational Health and Safety Specialty Group (2019) at the Society for Risk Analysis. She is a founding member of both WHWB and its US branch (WHWB-US) and has been on the ANSI Z10 Committee since its inception (2000). She has numerous publications and presentations. Mary is a Fellow of the AIHA and the recipient of the AIHA 2016 Alice Hamilton Award. Her book *Teen Health from Head to Toe: Exploring Issues and Risks* was published in 2020.

Dr. Vinka Longin Peš graduated from the Faculty of Law, University of Zagreb, Croatia, where she defended her doctoral thesis in the field of Civil Law. After the bar examination, in 1998, she worked in her law office until 2008. Since 2009, she was employed at the Croatian Health Insurance Institute for Occupational Health as the Head of the Occupational Health Insurance Department. After merging this institution with Croatian Health Insurance Fund, Vinka worked in the Department for Health Protection at Work, Analysis and Coordination of Litigation for Damages. Since 2020, she has been working as the head of the Project Department. She has published scientific and professional papers and participated in conferences, projects, and working groups at home and abroad.

Diana de Sousa Policarpo graduated in Law from the Faculty of Law of the University of Lisbon, Portugal; practiced in the small town of Marinha Grande (1993–2000); and joined the General Labour Inspectorate (current Authority for Working Conditions – ACT) in 2001. In 2012, she became the Head of the Division of Promotion and Evaluation of Programs and Studies of the Directorate of Services for the Promotion of Safety and Health at Work at ACT, Lisbon. Currently, Diana is Labour Inspector at ACT's Vila Franca de Xira Local Unit and is completing the second year of the Master's course in Management and Public Policies (2020–2021) at the Higher Institute of Social and Political Sciences of the University of Lisbon. Since 2015, she has been representing ACT in the coordination of the Mind Safety – Safety Matters! Project, supported by the National Agency ERASMUS+. This project aims to establish an interface between teacher education, professional training, and learning contexts in OSH. By supporting teachers' education, this project will help them to expand their knowledge, skills, and attitudes and provide the right tools to deal with OSH issues at school. She is currently Co-Chairman of the ENETOSH Steering Committee. She has taken part on several missions abroad representing the Project, ACT, and ENETOSH (2016–2020) facing different degrees of safety culture.

David S. Rodriguez Marin started his professional career in Eli Lilly and Company in Mexico City as an industrial hygienist for the manufacturing site. In 2004, David joined 3M Mexico as Technical Service Engineer for Occupational Health & Environmental Safety Division. From 2008 to 2012, David was appointed as Latin America Technical Affairs Manager for Occupational Health & Environmental Safety Division. Since 2012, he has been Project Manager of Innovare EHS, S.A. de C.V.; and Vice President in Analisis Ambiental, an accredited Industrial Hygiene Laboratory, based in Mexico City. David is the Vice President of Mexican Industrial Hygiene Association. He collaborates with several professional associations in Latin America focusing his efforts to develop the IH practice in the region. David leads the TLV and BEI translation to Spanish. Since 2015, he has been a member of the IOHA Board of Directors. He earned a Chemical Engineering BA degree from Universidad Iberoamericana, and he has a Diplomat in Environmental, Health and Safety by UNAM. David is a Certified Professional in Industrial Hygiene by the Mexican Council of Certification for Professionals in Risk Prevention.

Laurence Svirchev has a 40-year career in OHS. He has been a CIH since 1994. He earned an MA in Disaster and Emergency Management from Royal Roads University and a BSc in Occupational Health and Safety from the Montana School of Mines. Svirchev spent a decade in occupational epidemiology research at the BC Canada Cancer Agency (Canada), and 15 years at WorkSafeBC as an Occupational Hygiene

Officer and Fatalities Investigator. Since 2010, he has worked as Hygiene-Safety-Environment-Social Manager for Chinese international infrastructure construction companies in China, Indonesia, Fiji, and the Sahel region of Africa. In the years of COVID-19, he has been a COVID Compliance Officer in the film industry. Svirchev has authored and co-authored 17 peer-reviewed articles in the IH, disaster management, and geology literature. He is an active member of the AIHA and multiple ad hoc committees and coalitions promoting the health and safety of working people around the world.

Dr. Steven M. Thygerson is an Associate Professor in the Department of Public Health at Brigham Young University (BYU) teaching occupational health and toxicology courses for the past 13 years. Prior to his appointment at BYU, he worked for 9 years in various occupational and environmental health settings in the public and private sectors. Steve earned his Bachelor's degree in Zoology from BYU, a Master's degree from the University of Utah, and his PhD in Environmental Health from Colorado State University. He is a CIH. He has been a member of WHWB for 5 years mentoring other occupational hygienists and providing occupational health training worldwide. Those countries include Nepal, South Africa, Mozambique, Ethiopia, and Pakistan.

Timothy David Tregenza is Senior Network Manager at the EU-OSHA. After studying International Relations at Lancaster University, he became a labor inspector for the UK's Health and Safety Executive (HSE) in the North West of England. Following a secondment to EU-OSHA in Bilbao, Spain, as a detached national expert, he took up post there as a project manager before taking up his current role as senior network manager. Official: To learn more about EU-OSHA classifications, visit https://osha.europa.eu/en/ic.pdf.

Noel Tresider is a Fellow of the AIOH and a retired Certified Occupational Hygienist (AIOH); a retired CIH (ABIH); and a Visiting Fellow of the University of Wollongong. He was AIOH President (1998–1999), IOHA President (2011–2012), and an IOHA Board member (2006–2013), and received the IOHA Lifetime Achievement Award in 2014. He received the AIHA Yant Award in 2015. In 2018, he was honored by the Australian government with the Member – Order of Australia (AM). He became involved with the OHTA (2006–2015) and Board member (2015–2018). Tresider is a member of the Asian Network of Occupational Hygiene (ANOH); a board member of WHWB (2015–2017); and a member of the Hong Kong Occupational & Environmental Health Academy – International Advisory Group (2015–2016). He has more than 40 years' experience in the oil and chemical industries and has provided IH services to Mobil Oil affiliates in the Asia–Pacific region. More recently, he has provided IH services to other major oil companies in Australia and Qatar. He is currently employed as the Chemical Project Officer at Baker Heart and Diabetes Institute in Melbourne.

1 Foundations and Networks of Occupational Hygiene Capacity

Thomas P. Fuller
Illinois State University and the International
Occupational Hygiene Association

CONTENTS

1.1 INTRODUCTION

There are approximately 3,514,988,000 workers in the world over the age of 15 (ILO, 2020a). Each year 2.78 million workers will die on the job as a result of accidents and exposures to toxic chemical, physical, and biological agents (Hämäläinen, 2017). This amounts to five workers around the world dying every minute. Worldwide, there are around 340 million occupational accidents and 160 million victims of work-related illnesses annually (ILO, 2020b). And in many countries, these numbers are grossly underreported.

1

In the early part of the last century during the industrial revolution, a new profession was created to anticipate, recognize, evaluate, and control workplace hazards as a means to reduce work-related injury, illness, and death. This profession is called occupational/industrial hygiene.

Today there are approximately 20,000 occupational hygienists striving to make workplaces safer and healthier throughout the world. Yet despite this number, workers continue to be injured and die as the result of workplace hazards. The toll of workplace injury, illness, and death is indicative of the urgent need for additional occupational hygiene knowledge and person power globally. Almost every fifth grader knows the title "nurse" and generally knows what nurses do. But unfortunately, not everyone is as familiar with what an occupational hygienist does. Furthermore, the shift of basic manufacturing away from the United States due to neoliberal and anti-regulatory policies has reduced industry-based employment for occupational hygienists. This has resulted in universities struggling to attract enough students into the field of study.

In countries with more advanced economies, where more regulations have been promulgated that require employers to take certain actions to protect workers from workplace hazards, there tends to be more occupational hygiene capacity in terms of educational institutions that provide specialized coursework, and thus a larger pool of qualified and competent practicing professionals. Often in these same economically advanced nations, businesses and organizations have realized the clear financial benefits of having a safe and healthy workforce, fewer accidents that destroy infrastructure and disrupt business, and reduced insurance premiums related to both. The value of occupational hygiene capacity in these nations is fairly well understood to not all, but many, business and government organizations.

Approximately 8,000 occupational hygienists are in the United States; half of them are certified. The United States also has the most academic Occupational Hygiene (OH) programs with more than 75 schools offering degrees at all levels. Canada has the second largest number of industrial hygienists, with more than 600. China, the most populous country in the world, has proposed a fast track in its development of safety and health systems by quickly moving to a mandated management systems approach while trying to establish the professional resources to carry this out.

Globalization has shifted worldwide manufacturing from the economically advanced nations to Asia, India, Mexico, and other regions. As a result, more occupational hygiene professionals are dealing with international issues even if only limited to requirements for exports. It also has led to the growth of occupational hygiene outside the United States, along with some concepts such as the "Green Movement" and "Sustainable Development" from Europe and other places.

In many developing nations, the economic and social benefits of investment in occupational hygiene have been relatively underreported and are therefore less understood and utilized by governments and businesses. Not only is the creation of new health and safety workplace regulations a slow process, but businesses are less aware of the value of the practice of occupational hygiene in their profit models. In many economically developing countries (EDCs), and in nations without robust OH educational programs, the shortage of qualified occupational hygienists is even more alarming. As a result, employees in these countries suffer higher rates of workplace injuries and illnesses (Hämäläinen, 2009). Damage to valuable infrastructure, the environment, and public health also occurs in areas without adequate professional OH support.

Privatization of workers' compensation systems is another driver, potentially expanding interest in occupational hygiene globally (South America, for example). This provides additional financial incentives for establishments to reduce worker injuries and illness. This is important because many government safety and health inspection programs do not effectively penalize those with the highest injury and illness rates.

In recent years, in both developed and developing nations, several grass roots and nongovernmental professional organizations have arisen to fill the void in occupational hygiene capacity. This book is a discussion of many of the most predominant of those organizations, and those institutions that have had the greatest impact on building occupational hygiene capacity globally.

In understanding the importance of occupational hygienists and the role they play in worker health and well-being, many nonprofit organizations have been developed to directly address some of the more egregious problems in EDCs. Through an elaborate, but transparent network of collaboration and communication, numerous educational and research projects have been completed to assist EDCs in the development of OH capacity over the past few years. Documenting the stories of how these groups formed, and the work they continue to do, will be of interest to those currently practicing OH. New graduates entering the field, as well as experienced and emeritus professionals, who want to do philanthropic international work will also gain important insights.

This book outlines a broad variety of capacity-building projects in OH. It also reviews some detailed experiences of occupational hygienists working abroad, or working on special global OH topics.

This book may be of most interest and use to occupational hygiene professionals with several years of experience who are interested in extending and broadening their careers with philanthropic work abroad. Professionals who have spent their lives protecting workers from occupational hazards know the value of what they do and expanding their influence globally can help uplift workplace standards and economies. Learning about the nonprofit organizations, and what type of work they are doing, will help match the professional's interest with the needs of the organizations. Knowledge through networking and collaboration between these organizations and other professional and governmental organizations is an excellent way to begin to address the global need for occupational hygienists. This book can serve as a link between these organizations, and a one-stop source of information.

Sections 1.2 and 1.3 introduce two organizations, the American Industrial Hygiene Association (AIHA) and the International Occupational Hygiene Association (IOHA), that have played, and continue to play, a significant role in initiating many of the collaborations discussed in later chapters. Many of the members in the other organizations discussed in this book are also members of AIHA and IOHA. In fact, it would probably be difficult to find many practicing occupational hygienists who were not, at least by association of their national professional organizations, members of one of these two organizations.

1.2 AMERICAN INDUSTRIAL HYGIENE ASSOCIATION

The American Industrial Hygiene Association (AIHA) is an association of professionals committed to preserving and ensuring occupational and environmental health and safety in the workplace and community. AIHA was formed in 1939 as a nonprofit organization in the state of Illinois, USA. The overall objective of the organization is to help ensure that work-related occupational and environmental health and safety hazards are anticipated, evaluated, and eliminated or controlled.

The mission of AIHA is to empower professionals to apply the technology of occupational hygiene to protect all workers from hazards in performing their jobs. The ultimate vision of the organization is a world where all workers remain healthy and safe.

The actions of AIHA include promoting the profession and practice in industry, government, and the community. It works to expand the knowledge, competence, and credibility of its members practicing the profession. AIHA strives to provide an international forum for the exchange of information and ideas about occupational hygiene internationally.

Within the AIHA, work is performed primarily by volunteers who serve in a variety of subsets of the organization. Volunteers may serve the AIHA by working in one or more of seven Working Groups, three Special Interest Groups, five Professional Development Committees, four Internal Operations sections, four Advisory Groups, and three Task Forces. Each of these subgroups meets regularly to exchange information, advance new technologies, and promote the profession within the organization and more broadly to the community at large. A list of all the various groups is provided in Table 1.1.

TABLE 1.1
Listing of AIHA Volunteer Groups, 2020

Technical Committees

Aerosol Technology	Sampling and laboratory analysis
Biological Monitoring	Social concerns
Biosafety and Environmental Microbiology	Stewardship and sustainability
Cannabis Industry Health and Safety	Teen workplace health and safety
Communications and Training Methods	Toxicology
Computer Applications	Exposure assessment strategies
Confined Spaces	Government relations
Construction	Hazard prevention and engineering controls
Environmental Issues	Indoor environmental quality
Ergonomics	International affairs
Exposure Control Banding	Ionizing radiation
Occupational and Environmental Epidemiology	Laboratory health and safety
Protective Clothing and Equipment	Leadership and management
Real-Time Detection	Legal issues
Respiratory Protection	Noise
Risk	Non-ionizing radiation
Safety	

(Continued)

TABLE 1.1 (*Continued*)
Listing of AIHA Volunteer Groups, 2020

Working Groups

Healthcare	Nano- and advanced materials
Incident Preparedness and Response	Oil and gas
Mining	Opioids
Museum and Cultural Heritage Industry	

Special Interest Groups

Academic	Minority
Fellows	

Professional Development Committees

Career and Employment Services	Student and Early Career Professionals
Joint Industrial Hygiene Ethics Education	Women in IH
Mentoring and Professional Development	

Internal Operations

Conference Programs	Finance
Continuing Education	Publications

Advisory Groups

Academic	International
Content Portfolio	Strategic Technology Initiatives

Task Forces

Total Worker Health	Body of Knowledge Framework Project
Leading Health Metrics	Team

1.2.1 INTERNATIONAL AFFAIRS COMMITTEE

As one of the 33 technical committees promoting the AIHA mission, the International Affairs Committee (IAC) has a number of interrelated objectives. Fundamentally, the IAC goal is to expand the awareness of occupational hygiene principles and concepts and build capacity in regions of the world where it is most lacking. Residents from outside the United States and Canada can obtain an E-membership at a greatly reduced rate of only US$56 per year. This membership includes access to the *Synergist* magazine, members-only content, the online communication community, and other resources at reduced rates. E-members are also encouraged to serve on and lead AIHA committees.

Members of the IAC typically come from a broad range of countries around the world. During monthly meetings, topics of international concern are discussed. But it is also a place to raise questions and issues with individual countries and regions, to seek advice and guidance from members from other nations, and to build collaborative networks of communication and support.

The IAC is a place where members can expand awareness of international workplace issues. The group advocates for policies, standards, and initiatives to prevent exposure to hazardous working environments in all countries and all types of jobs or working conditions. Meetings are a place where needs are expressed, and challenges are met through discussion and development of collaborative projects. Sometimes these projects involve coordination and development of occupational hygiene training on various topics in a region. Sometimes projects involve scientific research on exposure to toxic agents and methods to control the associated hazards.

The IAC promotes discussion and understanding of important global occupational hygiene issues. Information is broadly disseminated to educational institutions, governmental and public health agencies, labor organizations, nongovernmental organizations, and employer organizations. Organizations with representatives who regularly participate and report to the IAC meetings include Workplace Health Without Borders (WHWB), the Maquiladora Health & Safety Support Network (MHSSN), the Developing World Outreach Initiative (DWOI), the Occupational Hygiene Training Association (OHTA), and the European Network Education and Training in Occupational Safety and Health (ENETOSH).

Over the years, the IAC meetings have become a springboard of collaborative projects between the members who participate from around the world, and the organizations supported by volunteers to conduct research, disseminate information, and provide training in remote regions of the globe.

> **Case:** In 2017, an IAC member became aware of a small college in Gaborone, Botswana who would like support with the development of their bachelor's degree in Occupational Safety and Health. This was brought up in an IAC meeting and discussion ensued. Through a collaborative process over a period of six months, and eventual agreements between OHTA and WHWB, an IAC member was able to teach the OHTA week-long W201 Basics of Occupational Hygiene course at the college. Since that time, as a result of these original IAC discussions, four additional OHTA week-long courses have been taught through regional contacts in eSwatini and South Africa.

Many of the other projects that were sprouted during discussions at monthly IAC meetings are discussed in later chapters of this book. Many of the collaborative activities in research, support, and training were initiated by contacts made through the IAC. These collaborations and networks then exist and expand over many years and in numerous global regions. Table 1.2 shows a historical list of IAC Past Presidents as a perspective of the longevity and reach of the group.

1.2.2 Emerging Economies Microgrant Subcommittee

For many years, the AIHA has supported numerous requests for funding various research and development projects. Requests were made to AIHA executives or staff, and the proposals were reviewed and approved by the AIHA Board of Directors on an ad hoc basis. In an effort to make the process more equitable and transparent, the IAC Emerging Economies Microgrant Subcommittee was created in 2018. The first

TABLE 1.2

International Affairs Committee – Past Chair List

John Henshaw	01/01/1996	01/01/1997
James Platner	01/01/1997	01/01/1998
Nick Yin	01/01/1998	01/01/1999
Hank Muranko	01/01/1999	01/01/2000
Keith Tait	01/01/2000	12/31/2000
Keith Tait	01/01/2001	12/31/2001
Paul Olson	06/01/1994	12/31/1999
Paul Olson	01/31/2000	12/31/2000
Arlene Golembiewski	01/01/2001	12/31/2001
Arlene Golembiewski	01/01/2002	12/31/2002
Andrew Cutz	01/01/2002	12/31/2002
Andrew Cutz	01/01/2003	05/12/2003
Roy Buchan	01/01/2003	05/12/2003
Stephen Reynolds	05/12/2003	12/31/2004
Stephen Reynolds	01/01/2005	12/31/2005
Stephen Chiusano	01/01/2005	12/31/2005
Stephen Chiusano	01/01/2006	12/31/2006
Wen Chen Liu	01/01/2007	12/31/2007
Brian Daly	01/01/2006	12/31/2006
Brian Daly	01/01/2007	12/31/2007
Mark Katchen	01/01/2008	06/30/2009
Marcos Da Silva	06/01/2009	06/01/2010
Douglas Dowis	06/01/2010	06/01/2011
Marianne Levitsky	06/01/2011	07/01/2012
Richard Hirsh	06/22/2012	06/04/2013
Muhammad Akram	07/02/2014	06/04/2015
Margaret Wan	06/04/2015	06/03/2016
Kul Garg	06/05/2016	06/04/2017
Thomas Fuller	06/05/2017	06/04/2018
Steven Thygerson	06/01/2018	06/01/2019

year, the Subcommittee received eight proposals to review and awarded funding to five of them for a total of approximately US$20,000. Funding included support to DWOI for various training initiatives in Lebanon and Nepal, MHSSN training programs in Bangladesh, and various research/development projects in Nepal. In 2020, the Microgrant process supported projects to evaluate artisanal mining practices in eastern Cameroon, training about silica hazards at regional conferences in Peru, and research to evaluate occupational hazards from e-waste in Vietnam for a total support amount of over US$23,000. Funding these projects on an ongoing basis, through a consistent, dependable, and transparent program, has been an excellent means to improve and expand occupational hygiene networks and to build capacity globally.

1.2.3 COLLABORATION WITH THE INTERNATIONAL
OCCUPATIONAL HYGIENE ASSOCIATION

As a member of the International Occupational Hygiene Association (IOHA), the AIHA maintains strong ties with 35 other national professional organizations globally. As a member of the organization, the AIHA has one seat on the IOHA board of directors. Responsibilities generally require attendance at biannual board meetings set in various member countries around the world. Board members also sit on or assume leadership roles on one of the five IOHA technical or administrative committees. AIHA IOHA board representatives are then in the position to both bring AIHA goals and initiatives to the attention of other IOHA board members and relay IOHA projects and strategies back to the members of the AIHA IAC during the monthly meetings and other regular communications.

The collaboration between IOHA member organizations and the other affiliations of other groups works to greatly expand the network and interconnections between dozens of national professional, nongovernmental, educational, governmental, and employer organizations who maintain representatives on the boards of IOHA and their own national professional organizations. In an ever-expanding web of connectivity, the capacity of occupational hygiene research, development, and education continues to expand.

1.3 INTERNATIONAL OCCUPATIONAL HYGIENE ASSOCIATION

The IOHA is a coalition of occupational hygiene organizations from around the globe. The primary mission of the IOHA is to enhance the international network of occupational hygiene organizations that promote, develop, and improve occupational hygiene worldwide, providing a safe and healthy working environment for all. It is felt that the key to building occupational hygiene awareness and capacity is through sharing and communicating as openly and broadly as possible.

The newly stated 2021–2025 IOHA strategies include the following list of measurable goals meant to help achieve the organization's mission:

* Expand nation IOHA membership (from 35 to 42)
* Increase worldwide membership (from 18,000 to 20,000)
* Increase the number of organizations participating in National Accreditation Recognition (NAR) (from 16 to 18)
* Improve committee productivity and collaboration
* Expand collaborations and partnerships with tripartite, nongovernmental, and educational organizations globally.

1.3.1 IOHA HISTORY

IOHA began in April 1987 at Nottingham, England, during a meeting between 15 occupational hygienists from various countries, to discuss the creation of an international association of the profession. Shortly after this, a second meeting was conducted in Montreal, Canada, with representatives from the professional associations of ten countries to clarify and agree upon the goals of the organization. By April of 1989, the organization was formalized with the completion and acceptance of the first governance documents, and the Association's bylaws were signed into effect by the board of directors (Fuller, 2020).

Today, IOHA has grown to include 35 national professional associations of occupational hygienists from around the world. Countries represented are listed in the white-background boxes in Figure 1.1.

1.3.2   IOHA COMMITTEES AND ACTIVITIES

1.3.2.1   Governance

IOHA is governed by representation of one board representative from each member nation professional organization. Each board representative has one vote on governance matters, and board members vote to select the organization Executive committee each year to consist of a President-Elect, President, and then Past President. A list of IOHA Presidents is shown in Table 1.3. The board also elects a Treasurer for a term of 2 years. New member nations' applications undergo a review of the organizations' bylaws and organizing goals and principles to ensure alignment and are accepted into IOHA by a vote of the existing board members. Member nations pay a small annual capitation fee to IOHA based on the population of their nation membership.

1.3.2.2   Communications

IOHA publishes a bimonthly newsletter called the Global Exposure Manager that highlights stories of activities in various member nation associations. Articles are also published that are of international or global interest. Many issues include information about regional conferences and expositions, upcoming training events or seminars, or new national or regional regulations on various occupational safety and health topics. Articles are also accepted and written about activities of international collaboration with global network affiliates and strategic partners in occupational hygiene.

IOHA also operates a website where members and the public can access information on timely topics including upcoming conferences and other events. This website also has information regarding IOHA membership and operations. The website can be accessed at www.ioha.net.

1.3.2.3   Stakeholder Relations

IOHA makes a concerted effort to reach out broadly to other like-minded organizations with interests toward worker health and safety. Memorandums of understanding (MoUs) have been developed to provide open access to information and materials from sister organizations and to foster collaboration and growth in the global occupational hygiene profession. As part of the 2021–2025 Strategic Plan, IOHA wants to expand collaborative activities with other global tripartite and professional organizations. Many of the collaborations are demonstrated through MoUs that outline

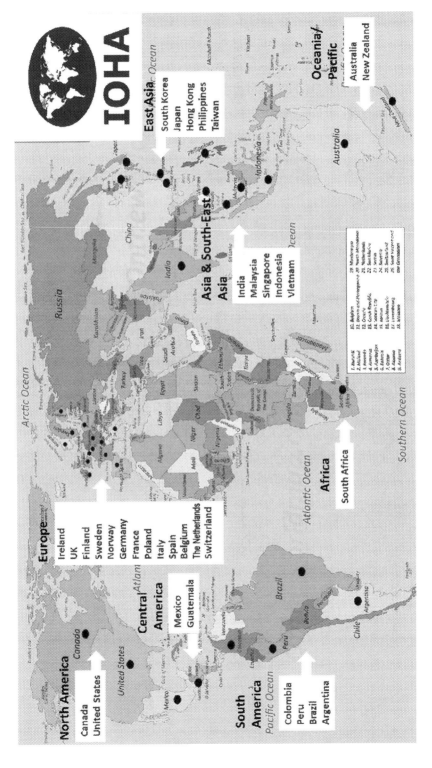

FIGURE 1.1 Countries with IOHA membership.

TABLE 1.3
List of IOHA Past Presidents

Year	President	Country
1988–1989	JS Lee	United States
1989–1990	PE de Silva	Australia
1990–1991	M Guillemin	Switzerland
1991–1992	V Riveira Rico	Spain
1992–1993	WH Krebs	United States
1993–1994	A Burdorf	The Netherlands
1994–1995	L Lillienberg	Sweden
1995–1996	RF Herrick	United States
1996–1997	TGE Gillanders	UK
1997–1998	R Viinanen	Finland
1998–1999	B Davies	Australia
1999–2000	P Oldershaw	UK
2000–2001	VE Rose	United States
2001–2002	V Leichnitz	Germany
2002–2003	D Zalk	United States
2003–2004	H Jackson	Australia
2004–2005	T Spee	The Netherlands
2005–2006	TW Tsin	Hong Kong
2006–2007	R Ferrie	South Africa
2007–2008	R Ferrie	South Africa
2008–2009	T Grumbles	United States
2009–2010	D Cottica	Italy
2010–2011	L Hamelin	Canada
2011–2012	N Tresider	Australia
2012–2013	J Naerheim	Norway
2013–2014	CC Chen	Taiwan
2014–2015	J Perkins	United States
2015–2016	K Niven	UK
2016–2017	DY Park	Korea
2017–2018	A Hiddinga	The Netherlands
2018–2019	P-J Jacobs	South Africa
2019–2020	R Leblanc	Canada
2020–2021	T Fuller	United States
2021–2022	NB Mydin	Malaysia

common goals and partnerships between the organizations. Table 1.4 provides a list of current MoUs between IOHA and other organizations.

1.3.2.4 National Accreditation Recognition Committee

The IOHA NAR committee was developed in 1999 to promote global respect for and recognition of Occupational Hygiene Certification programs which meet or exceed

TABLE 1.4

Past and Present IOHA Memorandum-of-Understanding Agreements

Organization	Dates in Effect
Workplace Health Without Borders (WHWB)	September 18, 2018–present
Occupational Hygiene Training Association (OHTA)	June 17, 2020–June 17, 2025
European Network Education and Training in Occupational Safety and Health (ENETOSH)	September 18, 2018–present
International Commission on Occupational Health (ICOH)	May 1, 2013–present
International Ergonomics Association (IEA)	May 1, 2013–present
International Society for Respiratory Protection (ISRP)	2015–present
World Health Organization (WHO)	2017–2019
European Platform for Professionals in Occupational Hygiene	November 2020–present

the "IOHA Model Certification Program". The process of accreditation is voluntary and aims to set a global standard for OH professional certification programs. It provides a measure of proof that national individual OH certification programs are sound and meet globally recognized best practice standards for OH professionals certified through their nation's certification programs.

Mutual recognition strives to ensure that all professionals certified at the OH/Industrial Hygiene (IH) level and whose OH certification program is accredited through IOHA are able to apply for and be recognized by any OH certification program as an OH/IH. Currently, 16 different national certification programs have been recognized by the IOHA NAR committee as shown in Table 1.5.

TABLE 1.5

National Accreditation Recognition Certifications

Australian Institute of Occupational Hygiene (AIOH)	Australia
British Occupational Hygiene Society (BOHS)	UK
Board for Global EHS Credentialing (BGC)	United States
Canadian Registration Board of Occupational Hygienists (CRBOH)	Canada
Dutch Occupational Hygiene Society (NVVA)	The Netherlands
French Occupational Hygiene Society (SOFHYT)	France
German Society for Occupational Hygiene (DGAH)	Germany
Hong Kong Institute of Occupational and Environmental Hygiene	Hong Kong
Institute of the Certification of the Figures of Prevention (ICFP)	Italy
Japan Association for Working Environment Measurement (JAWE)	Japan
Malaysian Industrial Hygiene Association (NYF)	Malaysia
Norwegian Occupational Hygiene Association (NYF)	Norway
Occupational and Environmental Health Society of Singapore (OEHS)	Singapore
Swedish Occupational and Environmental Certification Board (SOECB)	Sweden
Southern African Institute for Occupational Hygiene (SAIOH)	South Africa
Swiss Society of Occupational Hygiene (SSOH)	Switzerland

1.3.2.5 Education

The goals of the IOHA Education Committee (EC) are to expand awareness and understanding of the occupational hygiene profession through development of programs, curricula, and standards of practice through mutual discussion and consensus. The committee works to define, expand, and harmonize international occupational hygiene education, training, professional development, and recognition in countries where Occupational Safety and Health (OSH) capacity is needed most. The EC works with member organizations to identify new and emerging training and educational needs and work to develop new related training materials by coordinating collaborative support. Much of the Education Committee work is oriented toward the needs of EDCs.

The EC works to create or support international conferences or sessions on training and educational topics of interest and special concern to member organizations. Special efforts are undertaken with organizations external to IOHA in an attempt to expand awareness and understanding of the OH profession, and OH approaches to prevention of injuries and illnesses to workers. By collaborating on training and education projects with nongovernmental, governmental, and tripartite organizations to build consensus and standards of practice for OH training and education globally, the Education Committee helps to expand the availability of competent and educated professionals that are aware of recognized consensus standards of practice for OH globally.

The EC uses NAR foundational knowledge and competency requirements to build upon and further develop guidelines for OH training and educational curricula for bachelor- and masters-level college and university degree programs. Such guidelines and educational consensus standards are intended to be used as instructional models for future OH programs being developed or planned in developing countries, and for existing university programs interested in improving or expanding curricula in current OH programs or individual courses.

IOHA supports the activities of the OHTA, and the EC Chair is a member of the OHTA Board of Directors. IOHA supports and collaborates with OHTA on numerous projects. IOHA also collaborates with the ENETOSH to develop training and educational programs and approaches on occupational hygiene to reach students in technical colleges, secondary schools, and primary schools. Collaboration projects with ENETOSH in the past have included such activities as co-hosting education symposia at international conferences.

Currently, the Education Committee Chair represents IOHA on the International Labor Organization Global Coalition Special Development Group Task Force on "Promoting Decent Work and Productive Employment through Higher Education". This group, comprised of members from educational organizations from around the world, is working to develop a conceptual framework and a common understanding of key characteristics of education that contribute to improving the nature and environment of work. The goal is to identify educational initiatives, programs, and activities within higher education that can lead to better workplaces. As the work progresses, the group will focus on barriers to education and gaps between demand and the need for education and the available supply. As identified, the best means to expand education to meet content needs regionally will be disseminated to interested stakeholders.

The Education Committee Chair is also a Co-Chair of the planned Vision Zero Summit Special Session on Education/E-learning/Qualifications being scheduled for Japan in 2022. This education session will focus on diverse aspects associated with building OSH capacity by presenting campaigns, training programs, and projects that aim to raise OSH awareness and provide tools to strengthen OSH expertise at various technical and professional levels. The session will reflect existing and future challenges, as well as opportunities for OSH training. It will explore effective and meaningful ways to learn and teach OSH in both educational and occupational settings. The Summit will also cover strategies to provide global occupational safety and health training through advanced technologies in response to the global pandemic. The session will provide an open forum for sharing experiences and introducing innovative approaches to training including new digital technologies, and virtual reality systems.

1.3.2.6 Conferences

Three years after the inception of IOHA, they coordinated their first international conference in Brussels, Belgium. Since that time, technical conferences have been held every 2–3 years in a range of regions and nations. A list of IOHA conferences is shown in Table 1.6.

1.3.2.7 IOHA Awards

In 1997, the IOHA created a lifetime achievement award to honor the work of one of its members for their work in occupational hygiene and their service to the organization. Since that time, there have been ten recipients as shown in Table 1.7.

In 2018, the IOHA created the Collaboration Award to honor the work of occupational hygienists between nations and organizations sharing their ideas and efforts

TABLE 1.6
Location of IOHA Conferences, 1992–2020

Year	City	Country
1992	Brussels	Belgium
1994	Hong Kong	China
1997	Crans-Montana	Switzerland
2000	Cairns	Australia
2002	Bergen	Norway
2005	Pilanesberg National Park	South Africa
2008	Taipei	Taiwan
2010	Rome	Italy
2012	Kuala Lumpur	Malaysia
2015	London	UK
2018	Washington, DC	United States
2020 (October)[a]	Daegu	South Korea

[a] Postponed to 2021 and to be held virtually.

TABLE 1.7
IOHA Lifetime Achievement Awards

Year	Name	Country
1997	Jeffrey Lee	United States
2000	R. Jerry Sherwood	UK
2002	Bernice Goelzer	Brazil
2005	David Grantham	Australia
2008	Kurt Leichnitz	Germany
2010	Brian Davies	Australia
2010	Trevor Ogden	UK
2012	Michael Guillemin	Switzerland
2014	Noel Tresider	Australia
2018	Roger Alesbury	UK

in research and educational endeavors. The first recipients of this award, in 2018, were the Brick Kiln Committee of WHWB and the Global Fairness Initiative (GFI), through their joint research regarding hazards to brick kiln workers in Nepal. They worked together to establish a centre at Kathmandu University to collect data on sampling, results of analysis, medical information, and hazardous exposures in brick kilns as a basis for reducing hazardous exposures and improving worker health.

1.4 CONCLUSIONS

Although this book provides a good representation of many of the global OSH organizations collaborating and working together to expand the capacity of OSH, there remain numerous organizations that were not discussed. For example, IOHA is now an Affiliate Member of the International Commission on Occupational Health (ICOH). The ICOH is a nongovernmental organization founded in 1906 with a membership of around 2,000 professionals practicing in 105 countries. Many of the ICOH members are practicing nurses, physicians, and other health services providers with a focus on areas such as occupational medicine and the diagnosis and treatment of occupational illnesses and injuries. The ICOH collaborates broadly and is recognized by both the International Labor Organization and the World Health Organization. ICOH also maintains collaborations with the IOHA and other organizations described in later chapters of this book.

Surprisingly, at this time of the global pandemic of 2019–2021, perhaps because of a greater use of electronic communications platforms, we are seeing a burst of collaborative activities. And all indications are that these sudden relationships will continue to flourish moving forward. A new professional association created in 2017, the Guatemalan Association of Occupational Safety and Health (AGSSO), is already now a member of IOHA and involved in ongoing projects with members from other nations. The IOHA board representative from the Malaysian Industrial Hygiene Association (MIHA) was recently selected as the IOHA President-Elect. It could be assumed that this will lead to expanded occupational hygiene capacity in this region as networks of collaboration grow and strengthen.

The main goal of this book is to provide models for future collaborative projects in occupational hygiene by showing ongoing successes. The organizations and methods described here can be used as a structure and guideline for expanding networks further, perhaps into new geographic regions, perhaps into new topic areas of occupational hygiene. Readers are encouraged to reach out and collaborate.

REFERENCES

Fuller, T., The history of the International Occupational Hygiene Association, *Occ Health Southern Africa* (March/April, 2020) Vol. 26 No. 1 pp. 90–92.

Hämäläinen, J., Saarela, K., P., Takala, Global trend according to estimated number of occupational accidents and fatal work-related diseases at region and country level, *J Saf Res* (2009) 40 pp. 125–139.

Hämäläinen, P., Takala, J., Kiat, T., *Global Estimates of Occupational Accidents and Work-Related Illnesses* (2017), Workplace Safety and Health Institute, Singapore ISBN: 9789811148446.

ILO, Statistics on the working-age population and labour force, ILOSTAT Webpage, International Labor Organization, Geneva (2020a) Available at https://ilostat.ilo.org/topics/population-and-labour-force/ (accessed December 19, 2020).

ILO, World Statistics – The enormous burden of poor working conditions, International Labor Organization, Geneva (2020b) Available at https://www.ilo.org/moscow/areas-of-work/occupational-safety-and-health/WCMS_249278/lang--en/index.htm (accessed December 21, 2020).

2 Building OHS Capacity at the Grass Roots
A Case Study of the Maquiladora Health & Safety Support Network and Its Global Partners

Garrett D. Brown
Maquiladora Health & Safety Support Network

CONTENTS

2.1 INTRODUCTION

On the cusp of the implementation of the North American Free Trade Agreement (NAFTA) in 1993, occupational health and safety (OHS) professionals in Canada, Mexico, and the United States were concerned about the potentially adverse impact on workers' health and safety in the continent. If manufacturing plants moved from the United States and Canada to Mexico – as actually occurred – then Mexican workers would be exposed to increased workplace hazards but without effective government regulation to protect workers' safety and health. Mexico had, and has, OHS regulations equivalent to those in the United States and Canada, but near-zero enforcement due to lack of human, financial, and technical resources, and a pervasive culture of corruption in regulatory agencies (Brown, 2000).

In this context, the Maquiladora Health & Safety Support Network (MHSSN) was formed in October 1993 at a meeting of the Occupational Health and Safety Section of the American Public Health Association (APHA). Soon more than 300 OHS professionals had signed up to volunteer their time and expertise, all *pro bono*, for training, research, and technical assistance projects with worker and community organizations on the US–Mexico border.

MHSSN volunteers included industrial hygienists, toxicologists, epidemiologists, occupational physicians and nurses, and health educators. Among the professional associations that MHSSN volunteers belong to are the APHA, American Industrial Hygiene Association (AIHA), American Conference of Governmental Industrial Hygienists (ACGIH), National Fire Protection Association (NFPA), and the National Safety Council (NSC).

As global supply chains expanded in the 1990s, MHSSN projects also expanded in scope from the US–Mexico border to Central America, the Dominican Republic, Indonesia, China, and Bangladesh.

The 1998 mission statement of the Network declared that its goals were:

> To provide information, training and technical assistance about occupational safety and health to worker, community-based and human rights organizations in the developing world where global supply chains have production operations; to increase the knowledge and technical skills of grassroots and professional organizations seeking to protect the health and safety of workers and their communities; and to facilitate and strengthen ties between occupational health professionals in the developed world and worker and community-based organizations in developing countries in the interest of improving workplace health and safety throughout the global economy.

There are four key players in workplace OHS anywhere in the world: governments, employers, workers, and OHS professionals. The MHSSN has centered its activities on supporting and empowering workers and their organizations, as workers are the essential player everywhere with the least resources and support in the global OHS framework. The goal of the MHSSN is to provide the information and skills that workers and their communities need to effectively speak and act in their own name.

In the 25-year history of the Network, the volunteer staff established a directory of OHS volunteers with their skill set (which was provided to over 120 grassroots organizations), generated periodic "resource and readings lists" of key OHS materials, produced a newsletter (*Border/Line Health & Safety*), and created an ongoing website (http://mhssn.igc.org).

In addition to the donated time of professional volunteers, the Network applied for and received financial grants from professional associations (APHA, AIHA, NFPA), foundations (MacArthur, General Service, and Arca), and unions and labor rights organizations (United Steel Workers, International Labor Rights Fund, Worker Rights Consortium (WRC)) and funding from clothing and sports shoe brands for specific projects to establish factory health and safety committees (H&S committees).

The key to the Network's operations has been partnerships with OHS professionals on the one hand and with local networks on worker and community organizations at the grassroots level on the other hand. Since the Network's inception, the MHSSN has worked closely with the Labor Occupational Health Program (LOHP) at the University of California at Berkeley to conduct on-site trainings, generate user-friendly materials for low-literacy participants, and provide ongoing technical assistance to local organizations. The Network has also collaborated with the Labor Occupational Safety and Health program (LOSH) at UCLA. These partnerships have continued for more than 25 years.

During the first decade, the critical grassroots partnership was with the Coalition for Justice in the Maquiladoras (CJM), a border-wide network of four dozen community, women's rights, faith-based, and labor rights organizations on the US–Mexico border. In Asia, starting in 1999, the Network partnered with the Asia Monitor Resource Centre (AMRC) in Hong Kong and with the 25 member organizations in 17 countries of the Asian Network for the Rights of Occupational and Environmental Victims (ANROEV).

2.2 TRAININGS ON THE US–MEXICO BORDER

The first training on the border in Nuevo Laredo, Mexico, occurred in July 1993, prior to the formal founding of the MHSSN in October. Ten other "trainings of trainers" (ToTs) – primarily women workers in the *maquiladora* factories and community leaders in the surrounding neighborhoods – were conducted from 1994 to 2002, when drug cartel violence on the border made it too dangerous to continue. The worker participants of the ToT courses then went on to conduct their own workshops on OHS topics with co-workers and neighbors (Meredith & Brown, 1995; MHSSN newsletters, 1997–2012).

The 2-day trainings were usually held on the weekends – when both the participants and the volunteer instructors had time free from their jobs – in Tijuana, Ciudad Juarez, Matamoros, and other border cities. The instructors were Spanish-speaking Network volunteers, as well as staff members of LOHP, all of whom donated their time for the events.

The teaching methodology used consisted of "popular education" techniques which centered on interactive, participatory activities which drew on the experience and knowledge of the participants. Small-group activities included case studies, role-plays, and work groups. Training participants worked together to create hand-drawn "hazard maps" of their workplace, identifying exposures such as chemicals, heat, or noise. "Body maps" indicated body parts adversely impacted by exposures, and work processes were also developed by participant groups. In other exercises, small groups analyzed the safety data sheets of chemicals in use for health hazards and effective control measures or examined government regulations to determine the required protective measures and workers' rights under the law.

Written teaching materials included manuals and handouts with simplified language and graphics to explain key OHS concepts in a manner suitable for training participants with limited formal education. The easy-to-read training manuals of as much as 500 pages were developed over time, and included information on critical OHS topics such as toxicology, routes of entry, chemical hazards, noise, machine guarding, ergonomics, fire, and building safety. All training materials were translated into Spanish and, later on, into local languages in Asia.

By 2000, as a result of the Network's ToTs, a cohort of 45–50 women across the 2,000-mile border felt confident enough of their OHS knowledge and information to conduct their own workshops – usually for 1 or 2 hours – in their homes or in community locales with co-workers and/or community residents. Hundreds of workers and community residents on the US–Mexico border have participated in these trainings.

2.3 TRAININGS WITH UNIONS AND NONGOVERNMENTAL ORGANIZATIONS AROUND THE GLOBE

Based on the experience on the border, MHSSN volunteers have spent 20 years conducting similar ToT courses with trade unions and nongovernmental organizations (NGOs) around the world. The NGOs have included community-based, women's rights, human rights, faith-based, and labor rights organizations, all at the grassroots level.

The goal has always been to build the capacity of these base-level organizations so that their members understand basic OHS concepts and develop the confidence to act on the job or in the community to protect workers' health, safety, and legal rights.

The location and participants of selected ToTs conducted by MHSSN volunteers are listed in Table 2.1. Most of these trainings consisted of several days of classroom activities, utilizing popular education techniques, and also a "field day" in a working garment or sports shoe factory where participants could practice their hazard identification skills and see firsthand the available control measures for these hazards. The field day activities were facilitated by international brands, such as the Gap in Guatemala, Nike in Indonesia, and Nike, Reebok, and adidas in China.

TABLE 2.1

The Location and Participants of Selected ToTs Conducted by MHSSN Volunteers

Location	Year	Participants
Mexico City, Mexico	1995, 1996	Frente Autentico del Trabajo, independent union
Jakarta, Indonesia	2000, 2002	6 trade unions and 14 NGOs
Hong Kong, China	2000	2 trade unions and 8 NGOs based in Hong Kong but working in mainland China
Guatemala City, Guatemala	2001, 2003, 2004	3 trade unions and 12 NGOs, including independent monitors of garment factories in the six Central American countries
Manila, Philippines	2008	Participants of the ANROEV annual conference
Phnom Penh, Cambodia	2009	Participants of the ANROEV annual conference
Bandung, Indonesia	2010	Participants of the ANROEV annual conference
Santo Domingo, Dominican Republic	2010	Factory health and safety committee; union federation covering 7 factories
Dhaka, Bangladesh	2014	Council of 14 trade unions; inspection staff of Bangladesh Accord on Fire and Building Safety
Guadalajara, Mexico	2016	Labor rights and faith-based (Jesuit) organizations of electronics workers

As a result of the MHSSN trainings, the participating organizations began integrating OHS issues into their ongoing activities with workers in various industries. For many of the participants, these ToTs were the first OHS training they or their organizations had ever received. Many of these grassroots organizations continue to conduct OHS campaigns since the time of the MHSSN trainings in Asia and the Americas.

Local organizations have been able to use the native language manual and materials of the Network's ToT for their own educational activities. Some examples of these activities are provided here:

- In Indonesia, one union participant in the 2000 training adapted the ToT manual into an 80-page, Indonesian-language booklet, printed by the German Friedrich Ebert Foundation, with a run of 15,000 copies for distribution to union members.
- In China, Hong Kong-based NGOs established workers' centers and operated mobile vans in Guangdong Province with OHS curriculum and materials aimed at the millions of migrant workers in southern China.
- In Indonesia, organizations participating in the MHSSN trainings went on to organize a national campaign to ban asbestos use and led local efforts to improve working conditions for sports shoe workers.
- In Central America, increased OHS knowledge and skills were put to use by training participants in their inspections of the region's garment industry as independent monitoring organizations, and in national campaigns for enforcement of government regulations.
- In Mexico, community-based organizations in Guadalajara have conducted a series of educational activities on workplace safety with the primarily female workforce in the region's burgeoning electronics industry.

2.4   JOINT LABOR–MANAGEMENT H&S COMMITTEES

MHSSN volunteers have collaborated with international clothing and sports shoe brands in several projects to train the manager and worker members of joint H&S committees established on the factory level.

In Guangdong Province in China, MHSSN volunteers conducted a 4-day training at a Taiwanese-owned 30,000-worker sports shoe factory producing for adidas in Dongguan in 2001. There were 92 participants in the course – 25 workers and manager members from each of three factories (suppliers for adidas, Reebok, and Nike) and 17 participants from labor rights NGOs based in Hong Kong (Szudy, et al., 2003).

The Network's goal in the training was not only to help establish effective joint H&S committees in the three factories, but also to build the OHS capacity of the Hong Kong NGOs, who were and are monitoring working conditions in China's supply chain factories. The MHSSN has always declined to conduct management-only projects, insisting that all efforts include local capacity building for worker and community organizations.

Nine months after the Dongguan training, two of the Network's volunteer instructors visited each of the three factories to provide technical assistance and support for the joint committees which had been established in the factories. A photograph of one of a practice workplace inspection is shown in Figure 2.1. The three joint

FIGURE 2.1   Sports shoe workers in Dongguan, China, conducting a practice inspection to identify health and safety hazards in a 30,000-worker shoe factory in 2001. (Photo courtesy of Garrett Brown.)

committees – with 30, 60, and 100 members in the three plants of different sizes – also met together in February 2003 to exchange experiences, materials, and tips for running effective OHS programs at the factory level (Brown, 2003).

In the Dominican Republic in 2010 and 2011, MHSSN volunteers helped establish the safety program and train the joint labor–management H&S committee in a new "no sweat" garment factory in one of the country's "free trade zones." The Alta Gracia factory, owned and operated by the US-based Knights Apparel company, paid the workforce three times the average garment worker wage in the DR. The company recognized and bargained with a member-controlled trade union and asked the MHSSN to provide technical assistance in establishing the factory OHS program (Adler-Milstein & Kline, 2017).

Network volunteers conducted a pre-opening inspection of the facility in February 2010, another walk-around inspection and consultation with management in June 2010 when production began, and a series of follow-up visits in 2010 and 2011 to monitor progress. In June 2010, MHSSN volunteers conducted a 1-day training with the 30 members of the factory's joint H&S committee and a second 1-day training with workers of seven garment factories represented by the Free Trade Zone union, and labor rights groups from the Dominican Republic and Haiti (Brown, 2010b).

The operations of the Alta Gracia factory, which is now owned by Canada's Gildan corporation and still functions as the world's only "no sweatshop" garment factory, are highlighted by labor rights groups including the WRC in Washington, DC, as a model for other garment producers and other global supply chain factories (Kline & Soule, 2014).

2.5   OHS INITIATIVE FOR WORKERS AND COMMUNITY IN DHAKA, BANGLADESH

The work of MHSSN culminated in the establishment of the OHS Initiative for Workers and Community, a joint project of six NGOs in Dhaka, Bangladesh, in 2015.

The Network proposed the joint project to three labor organizations (a national union federation, a workers' center, and a labor research group), a women's rights group, a public health group, and an OHS organization in the aftermath of the April 2013 building collapse at the Rana Plaza building which killed 1,138 workers.

In October 2015, the six groups in Dhaka agreed to form the Initiative, and an ad hoc group of OHS organizations in California agreed to raise the operating budget for the first 3 years of the Initiative's operations. The "California Collaborative" includes the MHSSN, LOHP, LOSH, and Hesperian Health Guides. The Initiative began hiring staff in December 2016. The project has a seven-member Governing Board consisting of a representative of the six partner organizations in Dhaka, and one for the California Collaborative (California Collaborative, updates, 2016–2020).

In its first 3 years, the OHS Initiative trained 83 "master trainers" in three cohorts that were nominated by each of the six NGOs. The training participants – lead workers in garment factories as well as staff and supporters of the NGOs – go through a 20-day basic training, followed by a 5-day "refresher training." The instructors for this ToT include professors from local universities, staff of the country's Department of Labor and Fire Services, and leaders of the partner organizations.

In addition to essential OHS topics, the curriculum of the ToT includes gender-based violence, women's leadership, effective communication skills, and participation in factory H&S committees.

The master trainer graduates of the Initiative's ToT then conduct their own OHS workshops at the grassroots level with workers and community members in neighborhoods surrounding garment factories. To date, the 80+ trainers have completed 1- to 2-hour trainings in factories and communities with more than 4,000 people. Workshops with another 2,000 participants are planned by the end of 2020.

In addition, the OHS Initiative has put on annual 1-day seminars in Dhaka focused on gender issues in OHS and "youth and OHS" since 2018. The focus of activities to date has been on the garment industry (4 million workers), but the project plans to expand its outreach to workers in tanneries, construction, and ship-breaking.

The Initiative has also conducted periodic "trainers networking" meetings for the graduates of its ToTs so that they can exchange experiences, materials, and lessons learned with other peer-level trainers conducting grassroots workshops.

The first phase of the OHS Initiative will be completed in 2020, and a second phase with the goal of consolidating the grassroots-focused OHS project will begin with local leadership in fundraising in 2021.

2.6 LEGAL AND OHS POLICY WORK OF THE NETWORK

MHSSN volunteers have also offered their expertise to local worker organizations in the legal arena, and in developing OHS policy proposals for the profession as a whole.

Network members helped write and then testified at public hearings of three complaints filed under the "labor side agreement" of NAFTA that was designed to permit workers in one North American country to seek action in another country if their home country failed to protect their rights under existing national law.

MHSSN volunteers authored the health and safety sections of complaints filed by workers at Han Young de Mexico in Tijuana in 1998, and workers at Echlin/Dana Corporation near Mexico City also in 1998. In 2000, volunteers helped write the first all-H&S complaint filed under NAFTA on behalf of workers at the AutoTrim and CustomTrim plants in Valle Hermoso and Matamoros, Mexico (MHSSN, NAFTA Complaints, 1998–2000).

The Han Young workers were exposed to numerous safety hazards while working with malfunctioning cranes and poorly maintained welding equipment in the manufacture of truck chassis. The AutoTrim and CustomTrim workers had high levels of chemical exposures and repetitive motion injuries from producing leather-bound automotive steering wheels and gear shifts. The Echlin/Dana Corporation workers were exposed to asbestos hazards while manufacturing asbestos-containing brakes for motor vehicles.

MHSSN Coordinator Garrett Brown testified at US government hearings in San Diego on the Han Young complaint in February 1998 and on the AutoTrim/CustomTrim complaint in San Antonio in December 2000. Unfortunately, despite the overwhelming evidence of the Mexican government's failure to enforce its own H&S regulations, the NAFTA complaint process ended with agreements between the Mexican and US governments to hold public meetings, and to establish a government-only working group, on the OHS issues raised in the worker complaints.

In 2000, AIHA President Stephen Levine asked MHSSN Coordinator Garrett Brown to chair an AIHA Task Force on sweatshops and global supply chains. The Task Force, consisting of members of the association's International Affairs, Social Concerns, and Management committees, produced a White Paper and Board Position Statement, both approved by the AIHA Board in March 2001. The documents, containing 15 specific recommendations, were translated into Spanish and Portuguese.

In October 2007, the MHSSN responded to a request from the Mexican Miners union and the United Steel Workers union in the United States to assist miners at the historic Cananea copper mine in northern Sonora, Mexico. A multidisciplinary and multinational team of occupational health professionals went to the huge, 100-year-old open-pit mine to conduct medical screenings, gather work histories, and identify hazards during an inspection of the mine (Zubieta et al., 2009).

The Network team consisted of three occupational physicians, three industrial hygienists, a respiratory therapist, and an occupational health nurse, and the countries of origin for the team included Mexico, the United States, and Colombia. The team generated a report on the health and safety hazards at the facility and adverse health effects already detectable in the workforce and provided a series of recommendations to protect the lives and health of the miners.

The unions held a press conference in Mexico City, announcing the report's conclusions, and the Network volunteers met with a senior official of Mexico's Labor Department. As a result of the publicity, the mine's owner, Grupo Mexico, contracted with engineering and OHS consulting firms to address all of the findings of the MHSSN report and improve working conditions throughout the sprawling facility.

From 2011 onward, MHSSN volunteers conducted inspections at several garment plants in Central America on behalf of the WRC, a US-based labor rights NGO. The inspections were conducted at supplier facilities producing clothing for internationally known clothing brands supplying university-logo apparel to US colleges and work uniforms to the city governments of Los Angeles and San Francisco. The inspections confirmed worker reports of unsafe, unhealthy, and illegal working conditions at the supplier factories.

In 2016, MHSSN volunteers assisted the WRC in inspecting a 10,000-worker garment facility in southern Vietnam, which was producing clothing for US universities. The inspection documented illegal and unsafe working conditions, despite the fact that the plant had been repeatedly inspected by corporate social responsibility (CSR) monitors who failed to detect or correct the health hazards to workers (Brown, 2017b).

In all these cases, the brands' supplier factories inspected by Network volunteers were required to eliminate the identified workplace hazards and comply with national law and regulations.

Also in 2016, MHSSN Coordinator Garrett Brown assisted a lawsuit in Canada filed by Bangladeshi workers suing a Canadian company (Loblaws) and an international CSR monitor (Bureau Veritas) in the wake of the Rana Plaza disaster. Loblaws' clothing brand "Joe Fresh" was being produced at Rana Plaza at the time of the building collapse, and Bureau Veritas had made two inspections of the building on behalf of Loblaws and gave the structure a clean bill of health. More than 1,100 workers were killed in the collapse (Syed, 2018).

As an expert witness, Brown produced two affidavits for the Canadian law firm Rochon Genova and traveled to Toronto for deposition in 2017. The lawsuit, filed in April 2015, demonstrated how CSR monitoring has failed to protect workers in global supply chains. Despite the well-known history of building collapses in Bangladesh, Loblaws did not request – and Bureau Veritas did not offer – a structural evaluation of the building which had been renovated (three stories added) without permits to place industrial equipment and heavy electric generators on the upper floors. The situation exemplifies the "don't ask, don't tell" relationship between brands and CSR auditing companies. The case went to court in 2017, but the trial, appellate, and Supreme Court of Canada decided that the Bangladeshi survivors of Rana Plaza were subject to Bangladeshi – not Canadian – law and jurisdiction, and moreover, the statute of limitations had run out for the legal action (Doorey, 2019).

2.7 MHSSN CONTRIBUTIONS TO OHS SCIENTIFIC LITERATURE

Most of the MHSSN projects have resulted in written contributions to the OSH profession's scientific literature on working conditions in global supply chains, and the remedies needed to protect the lives of millions of workers around the globe. Starting at the time that NAFTA went into force, Network volunteers have written journal articles on working conditions in Mexico (Brown, 1999; Takaro et al., 1999), the weakness of NAFTA's H&S provisions (Brown, 2005b), and what is required in international trade treaties to protect workers' health, safety, and legal rights (Brown, 2005a).

Volunteers have published major articles on working conditions in global supply chains (Brown, 2002, 2010a, 2015a), including editing a special issue on OHS in China in 2003 (Brown, 2003; Brown & O'Rourke, 2003), the adverse impact of "lean manufacturing" on OHS (Brown & O'Rourke, 2007), and on working conditions in Mexican mines and Vietnamese garment factories. Since 2003, MHSSN Coordinator Garrett Brown has written a series of articles in trade and scientific journals on the failure of CSR programs of international corporations to protect workers in its global supply chains (O'Rourke & Brown, 2003; Brown, 2017a, 2018).

Network volunteers played a critical role in the development of a 530-page health and safety manual directed at workers in the developing world that was published as the "Workers' Guide to Health and Safety" by Hesperian Health Guides in 2015 (Jailer, et al., 2015). The MHSSN first proposed the book to Hesperian in 1997. A committee of 12 Network volunteers conducted a comprehensive needs assessment with worker organizations in the developing world, and the writing of the manual's text began in 2000. A $300,000 grant from the Rockefeller Foundation advanced the writing process, which was completed with internal funding from Hesperian.

2.8   MHSSN CONTRIBUTION TO OHS EDUCATION

In the fall of 2018, MHSSN volunteers coordinated a course on global occupational health at the School of Public Health at the University of California at Berkeley. The course, conducted again in the spring of 2020, was taught by UC Professor Kathie Hammond and MHSSN Coordinator Garrett Brown. Guest speakers in the course included Network volunteers and other professionals who presented their experience in protecting workers in the developing world in industries such as agriculture, pharmaceuticals, electronics, garment and sports shoes, energy (oil, gas, and coal), and mining. A unique aspect the of the UC Berkeley course was that the major project for the graduate students was to research specific OHS issues of concern to an NGO in the developing world. The students selected the organization and the topic and worked with their partner organization to define the issue and provide the group with information that would assist them in protecting their members and supporters in the developing world.

2.9   MHSSN IN THE OHS PROFESSIONAL COMMUNITY

Volunteers in the Network have been frequent presenters at conferences of major OHS professional organizations such as AIHA, ACGIH, NSC, and local affiliates of the associations in the United States and Canada since 1998. MHSSN Coordinator Garrett Brown has been an invited keynote speaker of OHS events at universities (University of British Columbia, University of Washington in Seattle, University of Oregon in Eugene, Pennsylvania State University, Stanford University, Global Labour University in Berlin); international OHS gatherings (World Safety Congress, Institution for Occupational Safety and Health conferences in the UK); and unions and labor rights organizations (United Steel Workers union and the global forum of the Clean Clothes Campaign).

The example of MHSSN's work has inspired other OHS professionals to organize similar efforts. In 2006, MHSSN members helped found and remain active members of the Developing World Outreach Initiative (DWOI) of the Northern California Section of AIHA. In 2011, Workplace Health Without Borders (WHWB) was launched by industrial hygienists in Canada, and now has chapters in the United States and other countries.

The work of the MHSSN was recognized in 2011 by the AIHA with its "Social Responsibility Award" presented at the annual conference for "the development and promotion of practical solutions to social responsibility issues related to OEHS."

2.10 CONCLUSION

The MHSSN is a small, voluntary network of OHS professionals who have donated their time and expertise for more than two decades to empower workers in global supply chains so that these workers can speak and act for themselves in improving working conditions in their place of work.

The work does not exist in a vacuum, however, and it has been affected by larger political and economic forces beyond the Network's control. During times of economic retrenchment on the US–Mexico border, for example, workers tend to reduce their health and safety demands on their employers so as to not lose their jobs. The commitment by international consumer brands sourcing from China to improve working conditions, for another example, tends to rise and fall depending on how much media and consumer attention is focused on "sweatshop" conditions in their supplier factories.

Despite these constraints, the MHSSN's work has set a useful example of how OHS professionals can use their skills and knowledge to strengthen the capacity of workers and their organizations in global supply chains. The Network has enhanced the workers' ability to understand and act on OHS principles to protect their own lives and those of their co-workers as well (Brown, 2015b). The combination of participatory, interactive teaching methods, accessible materials, follow-up, and ongoing technical assistance has made it possible for workers and their organizations in specific locations to increase the impact of their activities to protect their fundamental right to a safe and healthful workplace.

Over the last 25 years, the Network has learned the importance of being able to conduct follow-up trainings (one-off events yield little results) and ongoing technical assistance. Volunteers have worked to overcome the lack of accessible and understandable OHS materials appropriate for grassroots organizations by generating easy-to-read and easy-to-understand manuals, lesson plans, and materials that the workers can use for their own educational efforts.

The obstacles to putting on more trainings, developing more accessible materials, and providing ongoing professional-quality technical assistance are not the lack of interest on the part of Network volunteers or among base-level organizations of workers and community members. The major obstacle is the lack of financial resources for these efforts.

Given the ever-more harsh working conditions in global supply chains, and the failure of CSR and other top-down management systems approaches to protect workers, the further development of a worker-centered approach to improving global working conditions is essential. MHSSN volunteers have worked for years to add their "grain of sand" to the pile required to effectively protect the safety, health, and rights of workers around the world.

REFERENCES

Adler-Milstein, S., Kline, J.M., *Sewing hope: How one factory challenges the apparel industry's sweatshops*, University of California Press, 2017.

Brown, G.D., Failure to enforce safety laws threatens lives of Tijuana workers, *New Solutions* (1999), Vol. 9, No. 1, 119–123.

Brown, G.D., Double standards, U.S. manufacturers exploit lax occupational safety and health enforcement in Mexico's maquiladoras, *Multinational Monitor* (2000), Vol. 21, No. 11, 24–28.

Brown, G.D., The global threats to workers' health and safety on the job, *Social Justice* (2002), Vol. 29, No. 3, 12–25.

Brown, G.D., Taking steps to establish H&S committees in China, *The Synergist* (2003), Vol. 14, No. 3, 22–25.

Brown, G.D., Protecting workers' health and safety in the globalizing economy through international trade treaties, *International Journal of Occupational and Environmental Health* (2005a), Vol. 11, No. 2, 207–209.

Brown, G.D., Why NAFTA failed and what's needed to protect workers' health and safety in international trade treaties, *New Solutions* (2005b), Vol. 15, No. 2, 153–180.

Brown, G.D., Fashion kills: Industrial manslaughter in the global supply chain, *EHS Today* (September 2010a), 59–68. https://www.ehstoday.com/safety/article/21904714/fashion-kills-industrial-manslaughter-in-the-global-supply-chain.

Brown, G.D., More than an ad campaign: "No Sweat" in the Dominican Republic, *Industrial Safety & Hygiene News* (October 2010b), 49–50.

Brown, G.D., Bangladesh: Currently the worst, but possibly the future's best, *New Solutions* (2015a), Vol. 24, No. 4, 469–474.

Brown, G.D., Effective protection of workers' health and safety in global supply chains, *International Journal of Labour Research* (2015b), Vol. 7, No. 1–2, 35–53.

Brown, G.D., The corporate social responsibility mirage; Research shows minimal impact for the billions spent in the past 25 years, *Industrial Safety & Hygiene News* (May 2017a), https://www.ishn.com/articles/106349-the-corporate-social-responsibility-mirage.

Brown, G.D., Hansae Vietnam's garment factory: Latest example of how corporate social responsibility has failed to protect workers, *Journal of Occupational and Environmental Health* (2017b), Vol. 14, No. 8, 130–135.

Brown, G.D., Global capitalism undermines progress in workplace safety in Bangladesh's garment industry, *The Daily Star (Dhaka)*, June 22, 2018. https://www.thedailystar.net/star-weekend/longform/global-capitalism-undermines-progress-workplace-safety-bangladeshs-garment.

Brown, G.D., O'Rourke, D., The race to China and implications for global labor standards, *International Journal of Occupational and Environmental Health* (2003), Vol. 9, No. 3, 299–301.

Brown, G.D., O'Rourke, D., Lean manufacturing comes to China: A case study of its impact on workplace health and safety, *International Journal of Occupational and Environmental Health* (2007), Vol. 13, No. 3, 249–257.

California Collaborative, Updates on the OHS Initiative for Workers and Community (Dhaka, Bangladesh), 2016–2020, http://mhssn.igc.org/#bang.

Doorey, D.J., Canada's largest retailer, auditor not negligent in failing to protect workers at Rana Plaza, *International Labor Rights Case Law Journal* (2019), Vol. 5, 102–107.

Jailer, T., Lara-Meloy, M, Robbins, M., *Workers' Guide to Health and Safety*, Hesperian Health Guides, 2015.

Kline, J.M., Soule, E., Alta Gracia: Four years and counting, Georgetown University Reflective Engagement (2014), https://ignatiansolidarity.net/wp-content/uploads/2014/08/AltaGracia-LowRes-1.pdf.

Maquiladora Health & Safety Support Network, Border/Line Health and Safety newsletter, 1997–2012. http://mhssn.igc.org/newslist.htm.

Maquiladora Health & Safety Support Network, NAFTA complaints, 1998–2000. http://mhssn.igc.org/#mex.

Meredith, E., Brown, G.D. The Maquiladora Health and Safety Support Network: Case study of public health without borders, *Social Justice* (1995), Vol. 22, No 4, 85–87.

O'Rourke, D., Brown, G.D., Experiments in transforming the global workforce: Incentives and impediments to improving workplace conditions in China, *International Journal of Occupational and Environmental Health* (2003), Vol. 9, No. 3, 378–395.

Syed, F., What does a Canadian company owe to workers for overseas suppliers?, *Toronto Star*, May 6, 2018. https://www.pressreader.com/canada/toronto-star/20180506/281638190833831.

Szudy, B., O'Rourke, D., Brown, G.D., Developing an action-based health and safety training project in southern China, *International Journal of Occupational and Environmental Health* (2003), Vol. 9, No. 3, 357–367.

Takaro, T.K., Arroyo, M.G., Brown, G.D., Brumis, S.G., Knight, B.K. Community-based survey of maquiladora workers in Tijuana and Tecate, Mexico, *International Journal of Occupational and Environmental Health* (1999), Vol. 5, No. 4, 313–315.

Zubieta, I.X., Brown, G.D., Cohen, R.C., Medina, E., Cananea copper mine: An international effort to improve hazardous working conditions in Mexico, *International Journal of Occupational and Environmental Health* (2009), Vol. 15, No. 1, 14–20.

3 Workplace Health Without Borders

Engaging Volunteers to Improve Workplace Health and Safety in Underserved Regions around the Globe

Marianne Levitsky
Workplace Health Without Borders

CONTENTS

3.1   INTRODUCTION

For a decade, Workplace Health Without Borders (WHWB) has been coordinating efforts of volunteer professionals in occupational hygiene with education, project, and research needs in underserved populations around the world. This chapter describes how this organization began and grew. It also reviews its past and present projects and activities. Thoughts are provided on challenges and learnings.

> So many people have been involved in the creation and growth of WHWB that it would be impossible to credit them in this chapter without it reading like a telephone book (if anyone remembers what that is). I have therefore decided not to mention any names. While the information and history are factual to the best of my knowledge, there are so many initiatives under way that my knowledge is not always the best. The section on "what was learned" is my opinion alone.
>
> *Marianne Levitsky*

WHWB arose from meetings of occupational hygienists in 2009, including a meeting at the American Industrial Hygiene Conference and Exhibition (AIHce) in Toronto that year. Prior to that meeting, a proposed mission statement for "Industrial Hygienists Without Borders" was put forward the following problem statement and mission:

  I. Problem Statement: Over 90% of the world's workers do not have access to industrial hygienists. Therefore, nearly 3 billion workers will not be able to identify, control, and reduce work-related exposures to chemical agents and achieve prevention from acquiring work-related illnesses and diseases.
 II. Mission Statement: To engage the industrial hygiene profession in developing and disseminating the means for ensuring that the reduction of work-related illness and disease is attainable and available for all workers worldwide.

A small follow-up meeting was held at AIHce in 2010. In February 2011, a meeting of occupational hygienists was held in Mississauga, Ontario, Canada, at the offices of ECOH, a Canadian consulting company. The purpose of the meeting was to begin the process of launching the organization, building on the work conducted in 2009. The group proposed a Strategic Framework (WHWB, 2018), which evolved to become the founding document of the organization.

At a meeting at AIHce in Portland, Oregon, later in 2011, more than 40 attendees agreed to found the organization and asked that the Canadian group proceed to incorporate in Canada. WHWB, initially a "placeholder" name that stuck, was incorporated as a Canadian nonprofit organization in September 2011, and received status as a Canadian charitable organization in the spring of 2013.

Since WHWB's founding, national branches have been formed in the United States, the UK, and Australia. Each of these is incorporated in their respective countries and functions autonomously, but in close coordination with the WHWB (International) organization based in Canada.

In the 9 years since its founding, WHWB has focused on engaging the considerable energy and expertise of occupational health professionals in building occupational hygiene and health capacity in underserved areas of the world. It has remained an all-volunteer organization and has no paid staff. Its location of record is the ECOH office in Mississauga, Canada, where the organization was founded, but it functions as a virtual organization.

Since its founding, WHWB has grown to a network of more than 1,000 people worldwide, of whom nearly 700 have signed up as members. While most members are occupational/industrial hygienists, membership includes other occupational health professionals, most of whom are physicians.

The original Strategic Framework was modified in 2018 and now includes the following vision and mission statements and goals:

Vision
> A world in which workers, their families, and communities do not get ill because of their work.

Mission
> To prevent work-related disease and injury around the world through shared expertise, knowledge, and skills.

Goals
> To achieve WHWB's mission of preventing work-related disease and injury around the world by:

- Actively promoting the awareness and the practice of occupational health and occupational hygiene globally;
- Engaging volunteers in projects to improve workplace health for populations with limited access to occupational health and occupational hygiene expertise and services;
- Providing technical assistance, training, mentoring, and skill development to help workers and their representatives, employers, community groups, and public agencies develop the capacity and local infrastructure to manage and improve workplace health and safety conditions;
- Providing simple, practical guidance, tools, and methods for effective occupational health risk management;
- Working with organizations (such as government departments, non-governmental organizations [NGOs], professional societies, academic institutions, trade unions, industry, and others) who serve communities and workplaces particularly in developing countries, and populations with limited access to occupational health and occupational hygiene expertise and services, so as to integrate occupational health and occupational hygiene into their operations around the globe;
- Fostering collaboration among people and organizations working on common occupational health and occupational hygiene issues and problems; and
- Building knowledge and awareness about global occupational health and occupational hygiene needs among the general public, and within health care, academia, and occupational health communities.

The activities of WHWB can be grouped into two general domains – educational programs and projects.

3.2 EDUCATIONAL PROGRAMS

3.2.1 OCCUPATIONAL HYGIENE TRAINING ASSOCIATION (OHTA) COURSES

The initial impetus behind WHWB was the global need for professionals with occupational/industrial hygiene expertise. Only 16 countries have recognized criteria for certifying or recognizing occupational/industrial hygienists. In 2015, it was estimated that if the entire world were to have the same ratio of hygienists to workforce that exists in those 16 countries, the world would need about 45,000 more hygienists. From its inception, a major goal of WHWB was to help close this gap by offering training to those who work or want to work in occupational health/hygiene fields.

Most training programs offered by WHWB are courses of the OHTA, which is described in another chapter in this book. WHWB is an approved OHTA training provider and offers courses delivered by volunteers. While training by volunteers enables courses to be delivered at lower costs than courses given by paid providers, funds are needed to cover expenses such as trainers' travel, materials, and course venues.

All WHWB courses have been given in partnership with host organizations in the country where the course is delivered. To date, hosts have been universities and governmental or other public agencies. Universities which host the course usually aim to recover costs through student tuition; in one case, a grant was received from a private corporation to cover costs. Other host agencies have funded courses from their operational budgets.

In addition to the costs of trainers' travel and other operational costs, funds are needed to cover costs of the OHTA course examinations, which must be passed by the students in order to obtain a recognized certificate from OHTA. OHTA has recognized WHWB as an administrator of the examination for its W201 course, Basic Principles in Occupational Hygiene. Because WHWB volunteers develop and mark these examinations, they can be administered at lower cost per student, compared to fees charged by other test administrators. However, if WHWB delivers intermediate-level OHTA courses, the examinations must be administered by another approved test administrator, with higher per-student fees.

By the end of 2019, WHWB had delivered 11 OHTA courses in Tanzania, Zambia, South Africa, Botswana, Eswatini, and Vietnam. Additional non-OHTA courses were delivered in Tanzania, Mozambique, and India. Dr. Mosami Mogapi is shown in a small study session with students at OHTA W210 Basic Occupational Hygiene course at the Boitekanelo College in Gaborone, Botswana, in Figure 3.1.

A report on the OHTA course delivered in Kitwe, Zambia, in 2018 provides a flavor of experiences delivering these courses. This course was hosted by New Partnership for Africa's Development (NEPAD), an agency of the African Union, under its program to respond to the tuberculosis challenge in the southern African region.

The volunteer WHWB course instructors were occupational health professionals from South Africa. Most of the 20 students were employed at inspectorates of the

FIGURE 3.1 Students and professor at OHTA W201 course in Botswana in 2017. (Photo courtesy of Thomas P. Fuller.)

Ministries of Labor, Health, or Mining in Lesotho, Mozambique, Malawi, Zambia, and South Africa. One of the course instructors was bilingual in English and Portuguese, and provided assistance in Portuguese for the students from Mozambique. WHWB provided the W201 examination in Portuguese as well as English.

The course benefited from, as have a number of WHWB courses, participation or visits from senior government or agency officials. This course was opened by the Chief Occupational Hygienist of the Occupational Health and Safety Institute (OHSI) in Zambia, where the course was held. Other visits during the 5-day course were from the Director of the OHSI, the World Bank Senior Operations Officer in Zambia, and officials from the US Centers for Disease Control and Prevention (CDC) who were in Zambia at the time.

Students in this course were especially innovative and helpful in suggesting improvements and follow-up activities. These included recommendations for the following:

• The training to be spread out over a longer period – 1-week duration was deemed by most participants to be too short
• Additional training to build on the introductory course
• More time to be spent on practical sessions for demonstration and use of occupational hygiene measuring equipment
• Inclusion of a worksite visit as part of the training course

At the final session, participants were asked how they expected to apply their newly acquired occupational hygiene knowledge. Answers included the following:

- Creation of WhatsApp groups so participants could continue collaboration and networking
- Awareness raising and information dissemination at ministerial levels
- Advocating for purchase of sampling equipment
- Efforts to shift their home organizations toward a sustainable culture of occupational hygiene
- Training of other inspectorate staff in their home organizations
- Active communication with other ministries
- Training of employers and to impress upon them the importance of occupational hygiene capacity in workplaces.
- Sharing information with ventilation engineers in mines and process plants, to improve the management of these operations
- Endeavors to raise awareness and knowledge in the informal sector
- Developing a proposal to senior officials in their home countries to call for routine health surveillance and control measures in workplaces

3.2.2   NON-OHTA COURSES

WHWB is willing to develop and deliver customized courses at a host's request. As of the end of 2019, three such courses have been delivered:

- A 3-day course on Occupational Hygiene for Physicians delivered as part of a 3-week certificate program in occupational medicine at Maulana Azad University in Delhi, India
- A 5-day course on Basic Occupational Health and Safety at Eduardo Mondlane University in Mozambique
- A 5-day course on occupational hygiene and safety for Tanzania OSHA inspectors.

These courses were not standardized, but were developed by instructors based on their own knowledge and resources. The interest in this content indicates that there may be value in developing a standardized course that addresses basic safety as well as occupational health and hygiene.

3.2.3   COURSE EXAMINATIONS

Students and their employers value certification that indicates successful course completion. OHTA courses are especially valuable in this regard, as passing an OHTA examination confers a certificate that has international recognition. As an administrator of OHTA W201 examinations, WHWB has developed a detailed examination administration framework that has been approved by OHTA. Developing, transporting, managing, and marking examinations have proven to require intensive volunteer efforts, which WHWB regards as worthwhile because it enables students to take examinations at a much lower cost than that charged by other administrators.

Challenges have been encountered in administering and invigilating OHTA examinations. These are largely related to English examinations given to students for whom English is a second language. Many of these students had difficulty completing the examination in the allowed time, even though OHTA permits extra time for people sitting an examination in a second language.

3.3   MENTORSHIP PROGRAM

The WHWB Mentorship Program was established in 2012 with the intent of providing early-career Occupational Health and Safety (OHS) professionals with the opportunity to be mentored by more experienced occupational health/hygiene professionals. The need was identified because of the scarcity of experienced professionals in some areas of the world. Mentors and mentees are usually based in different geographical areas and meet by video conference. Assistance provided through meetings between mentors and mentees has included help with academic studies in occupational hygiene and medicine, practical discussions of day-to-day problems arising on the job, and demonstrations of how to use occupational hygiene instruments.

One mentee described her mentorship experience as "one of the most useful learning experiences I have ever had". She went on to describe the advantages she saw in the mentorship program, including the following:

- A very experienced occupational health professional sharing all his or her knowledge
- Sharing professional experiences with the mentor and analyzing practical cases
- Flexible timetable
- An opportunity for making friends all around the world.
- The possibility of helping other health and safety professionals by sharing knowledge

3.4   MONTHLY TELECONFERENCES

WHWB hosts monthly general membership teleconferences, each of which features a guest speaker who presents on a topic relevant to international occupational health. WHWB rotates times of meetings so that participants all over the world can join at least some of the time. Topics have covered a broad range of occupational health issues; recent presentations have covered diesel exhaust emissions, waste hazards, artisanal mining, nanotechnology, child labor, and hazards in the garment industry.

With the 2020 pandemic, WHWB started holding biweekly teleconferences on COVID-19. Topics have included respiratory protection, control banding, informal work, and protection of health care workers. These teleconferences have given participants an opportunity to discuss issues of shared concern following brief presentations.

3.5 INITIATIVES AND PROJECTS

3.5.1 Subgroup of Health Care Providers

While WHWB was begun by occupational hygienists, an increasing number of its members are physicians, nurses, and other health care providers. With this in mind, an occupational physician on the WHWB Board spearheaded formation of a sub-group of health care providers, with the aim of using collective expertise, training, and education skills to improve the availability of advanced occupational medicine to areas of the world where it may not be available. Intended activities include design and delivery of training programs, mentoring, research support, or hands-on delivery of services.

3.5.2 Issue-Specific Committees

WHWB members have formed two issue-specific committees that have brought together people from around the world with interest in a specific OHS issue. These have proven to be two of the most successful WHWB initiatives and, as models of collaborative efforts, point to the power of teamwork and matchmaking to accomplish OHS goals.

3.5.2.1 The Brick Kiln Committee

The Brick Kiln Committee has members who are active in addressing the hazards of brick making, especially in Nepal, Egypt, Tanzania, India, Bangladesh, and Pakistan. Members have collaborated and shared information on monitoring the exposure of brick workers in these countries to silica and other occupational hazards (Sanjel, 2018). The group also provided technical advice to an International Labour Organization (ILO) project on child labor in brick plants and to the development of standards for Better Brick Nepal. The committee has been successful in obtaining grants for a number of its initiatives, including development of a video on how to conduct air sampling for silica, delivery of training in Nepal, sampling for silica exposure of stonecutters in Tanzania, and establishing a Center for Environmental and Occupational Health in Nepal.

In 2019, the Brick Kiln Committee, together with Global Fairness Initiative, was the recipient of the first Collaboration Award given by the International Occupational Hygiene Association (IOHA).

3.5.2.2 The Waste Workers Occupational Safety and Health (WWOSH) Committee

The WWOSH Committee was spearheaded by researchers who are investigating hazards to waste pickers in Brazil, and brings together people interested in waste workers' health and safety around the world. Through its own page on the WHWB website, it collates research and information on waste workers, publicizes its projects, and seeks funding and support. Projects include OHS training for waste workers, acquisition and distribution of clay water filters, and assessment of metal exposure in children of waste workers. With the start of the 2020 pandemic, the WWOSH Committee has been actively collaborating with Women in Informal Employment: Globalizing and Organizing (WIEGO) in developing information resources to provide clear, accurate information on COVID-19 to waste workers and other members of the general public.

3.5.2.3 Agate and Gemstone Projects

In 2014, WHWB partnered with an Indian nonprofit organization, Peoples Training and Resource Centre, and the Occupational and Environmental Health Division of the University of Toronto Dalla Lana School of Public Health on a project to address silica exposure of cottage industry agate processors in Gujarat, India. The project was funded by Grand Challenges Canada and piloted prevention approaches to limiting dust exposure from agate, with the intent of adapting them for use in other low- and middle-income countries and a variety of industries. Ten worksites were recruited to participate in the pilot, to install and use custom-built local exhaust ventilation (LEV) systems to capture dusts at source. Other measures included educational tools that were developed and disseminated to workers, workers' family members, and neighbors of silica workers. A separate education tool was developed for primary care physicians treating workers who were experiencing symptoms of silicosis. A survey of workers indicated that knowledge of the hazard was high, but the workers' sense of efficacy in preventing illness was low (Falk, 2019). Workers were more likely to institute preventive measures when the work was performed in their own homes. Financial barriers to successful prevention practices were significant.

In 2018, WHWB partnered with the American Gem Trade Association and the universities of Delaware and Queensland on a related project to address exposure of gemstone workers in Rajasthan, India (WHWB, 2019a). Findings of this project indicated that silica exposure was much lower than among the agate processors in Gujarat. Wet methods to control exposure were largely successful.

3.5.2.4 Equipment Donations

WHWB has facilitated donations of industrial hygiene equipment such as sampling pumps, sound level meters, and noise dosimeters. In some cases, equipment suppliers have contributed refurbished equipment in good condition. Other donations of used equipment have come from individual members. Calibrating, maintaining, storing, and transporting equipment have proven challenging. A new initiative spearheaded by the Belgian Centre for Occupational Health (BECOH) shows promise to improve this element of WHWB activities, by providing well-maintained and calibrated equipment.

3.6 WHWB BRANCHES

Three branches of WHWB have been established to date in the United States, the UK, and Australia. There has also been interest in establishing a branch in Africa.

The branches are incorporated in their respective countries and function autonomously, with the organization incorporated in Canada serving an international coordination function. A document was agreed to among the four organizations setting out how the branches operate (WHWB, 2019b). This affirms the independence of each branch, but also commits the organizations to work cooperatively within the

WHWB Strategic Framework. As such, they all have similar mission and vision statements, but independently conduct projects and activities.

WHWB-Australia was formed in late 2018, and engages occupational health professionals in Australia in WHWB projects.

WHWB-UK includes members in Africa and the EU as well as the UK. Its activities include development of an e-learning program on silica awareness, and a competition to develop a low-cost LEV system. It has also actively supported the WHWB mentorship program and educational opportunities for mentees.

WHWB-US is based in Ann Arbor, Michigan. Its active projects include addressing hazards of nail salons and support for training programs in Bangladesh, Mozambique, and West Bengal, India. It also promotes student OHS projects through its student chapters.

3.7   PARTNERSHIPS

WHWB works largely through partnerships with key organizations in the OHS sphere. It has memoranda of understanding (MoUs) with the following:

The Occupational Hygiene Training Association (OHTA): The first organization with which WHWB signed an MoU, OHTA is a critical partner. Activities of the two organizations complement each other, and the MoU specifies that with respect to professional training, OHTA is responsible for program development, while WHWB's role is one of program delivery.

The Belgian Centre for Occupational Health (BECOH): BECOH is a nonprofit Belgian organization that operates an analytical laboratory for occupational exposure analysis. It offers pro bono analytical services for nonprofit groups and has been an invaluable partner in analyzing samples taken for WHWB and other projects, including silica samples to evaluate exposure of stonecutters in Tanzania and brick workers in Egypt, asbestos samples in India, and metals in biological samples from children of Brazilian waste pickers.

The International Occupational Hygiene Association (IOHA): An MoU between WHWB and IOHA was signed in September 2018 at the IOHA conference in Washington, DC. The MoU recognizes the alignment of the missions of the two organizations and sets forth the agreement to work together on initiatives such as research, volunteer recruitment, and mentoring.

The US National Institute for Occupational Safety and Health (NIOSH): A memorandum of collaboration between NIOSH and WHWB was signed by the Director of NIOSH and the President of WHWB in September 2019. It stated that the two organizations share a common mission to contribute to the prevention occupational disease and injury worldwide and that they plan to cooperate through initiatives such as training, mentoring, joint seminars, conference presentations, and scientific activities.

3.8 FUNDING

As an all-volunteer organization with donated head office space, WHWB has few administrative expenses. Therefore, resources needed for core operations are low. While part-time paid administrative assistance has been identified as a need for several years, funding has never been sufficient and reliable enough to engage staff on a sustained basis. Core funding has been raised through individual and corporate donations. Most projects have been funded through project-specific fundraising appeals and grants, although WHWB has been able to directly support a small number of low-cost projects.

3.9 WHAT WAS LEARNED?

As WHWB has evolved over nearly a decade, lessons learned point to the value of combining planning with the flexibility of being open to unexpected developments and opportunities. WHWB has embarked upon several strategic planning exercises over the years, but never adhered to a plan so rigidly that it could not seize new opportunities or respond rapidly to new developments. In many ways, the development of WHWB has been a lesson in serendipity.

A combination of the following features has enabled WHWB's growth, but also presented challenges:

- A large volunteer base that shares dedication and commitment to a unifying cause
- Being grounded in a profession whose small size fosters cohesiveness
- A worldwide geographical reach
- An all-volunteer, poorly funded core organization
- Virtually no bureaucracy or hierarchy
- A relatively open-ended mission that can encompass a variety of activities
- Synergy created by bringing together people in diverse cultures, career situations, and life stages, including the following:
  - Retired and semiretired members who can devote their experience, expertise, and time
  - Academic members who have access to research and university resources
  - Students and recent graduates who are enthusiastic, are tech-savvy, and can take on research projects
  - Employed members who can access the support (and sometimes resources) of their employers or own companies

A major factor underpinning WHWB's development has been the availability of inexpensive telecommunications resources that enable members all over the world to build a community and work closely together. Members who are separated by oceans and thousands of miles have built friendships and collaborations; sometimes, when meeting face-to-face for the first time, they are surprised to realize that they never previously met each other in person. Community building has also benefited from the nature of the occupational health and hygiene professions which, likely due to their small numbers, are especially cohesive and collaborative.

Bridging differences between cultures and regions has sometimes been a challenge for WHWB. One of the greatest challenges has been differentials in available technology and Internet connectivity, which can present barriers to participation in teleconferences and webinars.

WHWB members have had to learn the lessons of "cultural humility", a concept that emerged in the social services field that has been defined as "a process of self-reflection to understand personal and systemic biases and to develop and maintain respectful processes and relationships based on mutual trust...involves humbly acknowledging oneself as a learner when it comes to understanding another's experience". A minor but revealing example of barriers arising from cultural differences was a delay in administering an OHTA examination when students, coming from a culture where the family name is given first, were confused by the form that asked them to fill in their "first name" and "last name".

For members in developed countries, it is difficult at times to let go of accepted occupational hygiene approaches that may not be feasible or appropriate in some parts of the world. Examples of difficulties encountered include how to use wet methods in places with little water, set up LEV where there is little electric power, fit respirators where beards are a cultural or religious necessity, and apply exposure limits where work is conducted in workers' homes so that they are exposed to contaminants 24/7, along with their entire families including young children. In such circumstances, cultural (and professional) humility is a trait worth cultivating, as members learn to look to local wisdom and ingenuity for a fresh approach to occupational health problems.

A simultaneous advantage and difficulty has been the absence of a blueprint for establishing an organization like WHWB. While objectives and anticipated activities were set forth in WHWB's initial Strategic Framework, the path to accomplishing them was rarely clear, depending largely on serendipitous developments and connections between people prepared to take initiative and leap into the unknown. An example is provided by the unplanned introduction of a WHWB member to the head of BECOH, the nonprofit Belgian Centre for Occupational Health. This has led to a very productive partnership that enabled a number of exposure assessment projects that may have otherwise lacked the resources to proceed. A similar fortuitous connection was made when an early-career occupational hygienist in Tanzania became the mentee of the now-President of WHWB, volunteered to be his country's WHWB representative, and arranged the first OHTA course delivered by WHWB, which was given at Muhimbili University in Dar es Salaam in 2015.

The absence of bureaucracy and hierarchy has enabled WHWB and its branches to make rapid decisions and respond quickly to opportunities and developments, especially during crises. This capacity was seen, for example, in WHWB-US's initiative to provide occupational hygiene equipment when Hurricane Maria struck Puerto Rico. During the Ebola epidemic, WHWB sourced Personal Protective Equipment (PPE) for a hospital in Sierra Leone, and referred occupational hygienists to a company staffing African hospitals. During the COVID-19 pandemic, WHWB hosted frequent webinars and online discussions that not only addressed pressing questions for members but, more importantly, gave them a chance to directly interact with each other.

As an organization with limited funding, WHWB's approach relies largely on a "matchmaking" function – sparking connections between people and organizations that together can leverage resources – financial, human, or technical – to accomplish mutual goals. Sometimes this is as simple as introducing an OHS professional to someone who is seeking OHS advice, for example, a member who travels frequently to India offered to visit and provide technical assistance to a university professor seeking advice on laboratory safety. In other cases, travelers conveyed donated industrial hygiene instruments to universities in Tanzania and Pakistan who could use them in training programs. Other examples include pairing experienced academic members with younger researchers in developing countries who needed help with research and editing papers intended for publication. This has led to not only publications but also collaboration on further research.

More extensive initiatives have grown out of this kind of matchmaking, most notably the varied achievements of the Brick Kiln Committee and the WWOSH Committee who have successfully leveraged partnerships and participation of members from diverse geographical and professional arenas to obtain grants and carry out projects.

While the absence of paid staff and funding for a core organization has motivated major volunteer efforts, their downsides have included long timelines for getting things done and occasional lack of follow-through. As WHWB matures and looks ahead to its second decade, it recognizes the need to fill resource gaps that have persisted since inception.

As this is being written in the midst of the COVID-19 shutdown, WHWB also recognizes that plans for expanding its programs through on-site training and assistance may not unfold as envisioned just a few short months ago. Plans are evolving to offer more online training and programs in partnership with organizations around the world. WHWB's work in building a vibrant virtual network that spans the globe is standing it in good stead as its members continue to connect and collaborate across oceans.

REFERENCES

Falk, L., Bozek, P., Ceolin, L., Patel, J., Cole, D., Malik, O., Levitsky, M., Sobers, M., Reducing agate dust exposure in Khambhat, India: Protective practices, barriers, and opportunities, *J Occup Health* (2019) 61:442–452. https://doi.org/10.1002/1348-9585.12067.

Sanjel, S., Khanal, S., Thygerson, S., Carter, W., Johnston, J., Joshi, S., Exposure to respirable silica among clay brick workers in Kathmandu valley, Nepal, *Arch Environ Occup Health* (2018) 73(6), 347–350, DOI: 10.1080/19338244.2017.1420031

WHWB, Revised WHWB Strategic Framework (2018) http://www.whwb.org/wp-content/uploads/2018/03/Strategic-Framework-WHWB-Intl.pdf accessed June 23, 2020.

WHWB, WHWB Report: Jaipur Project: Coloured Gemstones Processing a Pilot Study for Worker Health Issues and Interventions (March 18, 2019) https://static1.squarespace.com/static/568a3352e0327c02e38c3369/t/5d2062f3397b46000194c322/1562403576448/WHWB+Jaipur+report+8mar2019+final+%281%29.pdf accessed July 12, 2020.

WHWB, How branches operate, (2019) http://www.whwb.org/wp-content/uploads/2019/01/How-WHWB-Branches-Operate-1.pdf accessed July 12, 2020.

4 Promotion of Occupational and Environmental Hygiene through Education
A Case Study of the Occupational Hygiene Training Association (OHTA)

Chris Laszcz-Davis
The Environmental Quality Organization, LLC.

Roger Alesbury
Consultant

Zack Mansdorf
Consultant

Nancy M. McClellan
Occupational Health Management, PLLC.

Noel Tresider
Petroch Services, Pty. Ltd.

CONTENTS

4.1 INTRODUCTION

Occupational-related injuries and deaths continue to be a widespread international problem, especially in the developing world, even though the problem has been well recognized for decades. A significant contributor to this problem is the lack of occupational health resources in much of the less developed world. At a 1994 World Health Organization (WHO) meeting in Beijing, China, it was stated that in many developing and newly industrialized countries, no more than 5%–10% of the working population (and in several industrialized countries, less than 20%–50%) have access to competent occupational health and safety (OHS) resources in spite of the evident needs (WHO, 1994).

A key element of this is the lack of adequately trained occupational hygienists who can anticipate, recognize, evaluate, and control occupational hazards. Based on this pressing need, the Occupational Hygiene Training Association (OHTA) was formed. OHTA provides an internationally accepted training and qualification framework which is a stepping-stone for developing professionals in their early careers. The training materials focus on practical "hands-on" aspects of occupational and environmental hygiene. The emphasis is on teaching the practicalities involved with the identification, assessment, monitoring, and control of workplace hazards in practical situations.

Quality, peer-reviewed teaching packages/modules available free to download at www.OHlearning.com are translated or suitable for translation into local languages, and can be used by a variety of institutions across the world. Although anyone can download and use the modules, in order to maintain quality and consistency, courses that lead to the international qualification are run by OHTA-Approved Training Providers (ATPs) and their organizations which have at least one professionally qualified occupational hygienist and have demonstrated the ability to provide quality teaching.

These modules provide a means for delivering consistent, quality training on core aspects of occupational and environmental hygiene that complement and enhance existing training opportunities available in academia and in many industrial organizations. They have been designed for early-career trainees to bridge the gap between the principles-level courses and master's-level programs. In the first 10 years of

FIGURE 4.1 Thermal environments course breakout session in Johannesburg. (Photo courtesy of Thomas P. Fuller.)

operation (2010–2019 inclusive), there were 190,000 OHLearning.com users, 10,037 examination candidates, and 1,162 courses in more than 50 countries.

More recent OHTA initiatives include broadened collaborations with a number of organizations, including the International Occupational Hygiene Association (IOHA), the International Commission on Occupational Health (ICOH), the WHO, the International Labour Organization (ILO), Workplace Health Without Borders (WHWB), and the National Institute for Occupational Safety and Health (NIOSH), to name a few. New course and module opportunities developed by OHTA include distance learning, a new focus on emerging health hazards, a renewed focus on age-old hazards in new applications, and expansion into safety-related issues which often accompany occupational and environmental health issues. Figure 4.1 shows students in a problem-solving breakout session at OHTA W502 on the thermal environment sponsored and taught by WHWB in 2019 in Johannesburg.

From early ideas in 2005, OHTA has evolved to become a significant global force in the training and development of the occupational and environmental hygiene profession. Throughout its development and evolution, OHTA has focused on identifying gaps in the resources available to develop occupational and environmental hygiene expertise. Its role in the global occupational hygiene community is as a resource that can be used by others to help with the common goal of improving worker health. It has achieved this through a strategy of collaboration with established professional bodies, aiming to enhance capability and create a career ladder for entry to the profession. The entry-level and intermediate course structure provides an accessible pathway for employers and students to progressively build occupational hygiene capability and improve worker health protection around the world.

Its structure is deliberately designed to be responsive to the views of the established global occupational and environmental hygiene community while retaining a razor-like focus on the needs of businesses and others to build occupational hygiene capability. A comprehensive history of the origins of OHTA is a paper by Alesbury and Bailey (2013).

4.2   THE NEED FOR OCCUPATIONAL HYGIENE TRAINING

4.2.1   OCCUPATIONAL INJURIES AND ILLNESSES

If we look at statistics alone, we already know that more people die each year from occupational injuries and diseases than from other major causes that are much more visible and increasing (armed conflict and violence, HIV/AIDS, and road traffic).

The ILO and WHO statistics reflect the following global fatalities (ILO, 2013):

- 2.78 million workers die each year from workplace causes, with 2.34 million (or 84%) of these dying from occupational diseases alone. This amounts to 6,411 deaths/day from occupational disease alone.
- By comparison, 381,000 (or 16%) of these 2.78 million workers die from occupational injuries alone.

These official statistics do not reflect other growing trends, such as the following:

- Environmental impacts of workplace agents.
- Environmental impacts of and diseases exacerbated by workplace agents (e.g., silica–TB; asbestos–smoking).
- Diseases that may be spread in workplace settings or which may require changes to working practices (e.g., COVID-19)
- Blurred lines between workplace, home, and community; exposed family, especially young and old vulnerable members.
- Workers in the developing economies who are not employed in formal sectors, with employment in the informal sector reaching 70%.
- Few medical facilities and treatment centers in emerging economies.
- Nonexistent public health registries for major illness and industry types.
- The reality that, while fatal illnesses outnumber the injuries, it is still "injuries" which are studied in detail, not "illnesses."
- The shortage of scientific skills (physicians and nurses, emergency response personnel, virologists, epidemiologists, public health experts, biochemists, bioengineers, occupational hygienists, and safety professionals, to mention a few).

In 2014, Lucchini and London stated that global OHS must be an international development priority (Lucchini and London, 2014). The reasons are compelling – economic globalization is leading to an increased OHS and public health gap. In developing countries, the absence of OHS infrastructure amplifies public health and development problems. Typically, existing occupational health institutions are underfunded. Additionally, no more than 5%–10% of workers in developing countries have access to skilled OHS practitioners. Economists generally assume (shortsightedly) that OHS, emergency preparedness, response and recovery, and public health are later developments in the social maturity curve and should normally be undertaken once the economy is strong enough to absorb the additional expenses required by preventive action.

As Nelson Mandela once said wisely, "Education is the most powerful weapon which you can use to change the world" (Mandela, 2017). That brings us to the OHTA and its inception.

4.3 OHTA HISTORY – INCEPTION TO FORMALIZATION

In **early 2005**, a number of senior occupational hygiene professionals working in multinational businesses shared their experiences regarding the challenges of finding qualified occupational hygienists. With many multinationals developing new facilities, particularly in extractive, pharmaceutical, and chemical industries, the challenges were particularly acute in developing countries. In addition to their corporate responsibilities to ensure safe and healthy working conditions, increased focus on sustainability and compliance required skilled risk assessment and control of the risks. However, in many parts of the world, there were few qualified occupational hygienists and limited understanding of the skills required to identify, assess, and control risks to health in the workplace.

In **August 2005**, this group met formally for the first time to share their experiences and look for ways to meet requirements for worker health protection as industrializing economies developed their resource extraction and manufacturing processes. Their central belief was that the same standard of control of health and safety risks should be employed wherever the operation might be in the world. This would avoid repeating some of the errors and tragic cases of ill health that arose from the industrial revolution of the nineteenth and twentieth centuries.

For various cultural and economic reasons in developing countries, there was often a lower expectation of OHS performance and limited local expertise. While addressing this concern, it was recognized that a major factor would be a need to work together to raise the general level of awareness and expertise. The aim was to create sustainable, local provision of sound occupational hygiene practice at the intermediate, "boots on the ground" level.

Over the next couple of years, this group of senior professionals met numerous times and engaged with the global occupational hygiene community to share experiences and debate ideas to address the challenge. At that time, there were members of the community who could not appreciate the need and believed demand could be met by sending trainees to established degree programs, most of which were in the United States. However, in practice, this approach posed a classic "chicken-and-egg" situation. Industry business leaders struggled with the significant financial outlay required to expatriate employees for 1 or 2 years while funding salary, tuition costs, and travel and accommodation.

Subsequently, an emerging view surfaced in support of a modular system of training, delivered locally and focusing on practical aspects of occupational hygiene at an intermediate level which could be taken back to workplaces and put to immediate use to improve health and safety. Having different levels of training would provide a step-by-step approach to building capability, thus helping to progressively develop skills and capabilities. The skills learned at the intermediate level could be put to immediate use in helping to improve workplace health risk management, as well as serving as a foundation for those who might have the aspiration and skills to study at a higher level and progress to full professional accreditation.

This culminated with the group preparing a position paper addressing the challenges and suggestions for the global community to move this concept forward (Alesbury, 2006). This was circulated widely in **October 2006**. Meanwhile, this group arranged for a pilot modular training course to be undertaken at the University of Wollongong, Australia.

After a process of testing this pilot course in October 2006, a model evolved based on a series of 1-week, face-to face instruction with a series of core modules written to cover core occupational and environmental hygiene technical and scientific practice (absent the regulatory requirements of any one country). The courses were initially developed and written under contract with funding provided by industry leaders BP and GSK. The materials were peer-reviewed and piloted before being adopted for widespread use. Each module included a 5-day program, student manual, slide packs, teaching guide, guidance for conducting practical hands-on sessions, resources for syndicate studies, and self-assessment questions for each day.

Later that same year, discussions were held with the IOHA, its member professional national associations, and examining boards to bring them up to speed with the developments and opportunities.

Over the next few years, a range of courses was developed. Those that are now still in use deliver practical, hands-on training in technical aspects of worker health protection. There is an integral, consistent student assessment process, meaning that the qualifications are transferable across the world. The modules are designed to allow local customs, practices, and regulations to be integrated "in country" by ATPs.

During the running of the various pilot modules, the British Occupational Hygiene Society (BOHS) developed the structure of the examination system in order to independently test that each student had a good understanding of the course material. The trialing of different examination question formats led to the firm conclusion that the use of short-answer questions was the fairest way to test understanding at the intermediate level, particularly bearing in mind the diverse cultural and educational backgrounds of students from around the world. The written examination is combined with a formative assessment of the required practical skills for the module concerned. This examination system was also endorsed by consultation with all of the main occupational hygiene certification bodies from around the globe. In addition, these certification bodies signed an important memorandum of agreement which recognizes the International Certificate in Occupational Hygiene (ICertOH) as a single global certification for occupational hygienists at the intermediate level.

In the first few years, a small volunteer steering group progressed plans. This included some of the original cross-industry group, as well as contributors from the American Industrial Hygiene Association (AIHA), the Australian Institute of Occupational Hygienists (AIOH), the BOHS, the IOHA, and the American Board of Industrial Hygiene (ABIH). There was a need for a more formal process and organization to take the scheme forward – holding true to the original concept but enabling rapid expansion in use without excessive cost or bureaucracy. Although the original team had progressed things to this point, there was a desire to hand this over to existing occupational and environmental hygiene organizations, to be an integral part of the global structure. Several options were explored with it being taken over by one or more national professional associations or IOHA. However, concerns about liabilities and a desired ability to move speedily meant that a new organization was needed, the OHTA.

In 2007, presentations were made to the boards of both AIHA and ABIH. In response to the presentation, ABIH formed its "Futures Committee" to specifically evaluate the OHTA modular training concept and means by which to support the initiative. The committee evaluation included a demand assessment, feasibility review, and recommendations for a path forward. After final presentation to the full ABIH Board of Directors and Executive Director, support for OHTA was agreed upon and a memorandum of understanding (MoU) was signed with OHTA in 2010 as a supporting (vs. voting) organization. ABIH clearly recognized OHTA as a viable means to grow global occupational hygiene competency.

In **2009**, OHTA was established as a UK Registered Company Limited by Guarantee to provide a legal entity and ownership for the training materials. It was set up to meet a global demand for practical occupational hygiene training as a non-profit voluntary organization, visible through its website. This organization, working under the guidance of IOHA which, together with AIOH and BOHS, jointly funded the development of a website (*www.OHlearning.com*) which was launched in **May 2010** at the American Industrial Hygiene Conference and Exposition (AIHce), where newly developed international training modules would be posted for free, open access. In the first 6 months of the website going live (**2009**), there were over 13,000 visits to the site from individuals in 116 countries, and 28 ATPs offering courses around the world.

This learning portal also provides links to participating IOHA professional member organizations, examining boards, and ATPs. Links enable site users to identify where and when courses are being offered.

Working with BOHS and others, a process was developed to appoint ATPs to run courses. These ATPs must meet certain criteria, including the requirement that the course director hold a professional qualification under the IOHA National Accreditation Recognition (NAR) program. ATPs run their courses on a chargeable basis.

To enable students to achieve recognized qualifications, OHTA signed an agreement with BOHS to provide examinations and award certificates for the OHTA International Training and Qualifications Framework. Although any organization can use OHTA materials and run courses, only OHTA ATPs can offer the BOHS examinations and awards under this framework. Those achieving six passes in a specified range of courses are eligible for the OHTA ICertOH awarded by BOHS. This award is an intermediate-technical-level qualification that may also be used as a stepping-stone to professional awards offered by professional organizations in the IOHA NAR scheme.

The principle adopted around the world was to share expertise and develop materials that could raise the standards of occupational and environmental hygiene practice on a global scale, with particular emphasis on developing economies. The material developed would be free of charge and "open access" to stimulate local development of the people, skills, and practices necessary to protect worker health. The need was to start from a basic level – developing skills at a technician level – on the expectation that this would stimulate higher level study and development – ultimately training to the level required for qualification to an IOHA NAR recognized standard. The opportunity for employers to "test the water" by sending an employee on a 1-week local course was a more feasible option to secure business buy-in. This local opportunity is especially important in economies that do not have readily available university programs for occupational and environmental hygiene, which includes most of the world. Experience subsequently demonstrated the value created when employees went back to their workplaces and were able to implement what they had learned, making approval of subsequent training more straightforward. Both employees and their managers benefited from the incremental, modular approach.

To this point in time, all development had been funded by donations, grants, or direct commissioning of course development by multinational companies. However, in 2012 it became clear that to ensure financially viability and sustainability, a different business model was required to provide core funding. This is now provided by a modest levy collected by BOHS on behalf of OHTA when students sit for the examinations. Prior to this, OHTA was run on an entirely voluntary basis, but this core funding enabled OHTA to sign an agreement with BOHS in 2014 to provide administrative and business support and the services of a Development Manager.

In 2014, there was a change to the OHTA structure and Articles of Association. For an interim period of 2 years, the OHTA Board included the Chief Executives of AIHA, AIOH, and BOHS as well as an IOHA representative. However, this changed in 2016 when OHTA Articles were again amended to enable OHTA to register as a charity in the UK, thus providing taxation and other benefits. The revised Articles of Association also made provision for national professional occupational hygiene associations (and other organizations that shared similar aims) to sign an MoU to become voting members of OHTA. Currently, the 27 "members" of OHTA with voting rights are WHWB and most IOHA member organizations (national occupational hygiene organizations). In effect, these "members" are the sponsors of the organization. This change resulted in the Chief Executives of AIHA, BOHS, and AIOH no longer holding formal seats on the OHTA Board.

Governance today is provided by a board of globally represented volunteers selected for their expertise, experience, and leadership with approval for their appointment by vote at an annual general meeting (AGM). In addition, a globally represented Advisory Committee was appointed by the OHTA Board to help the Board accomplish its mission and to provide a calibration framework for OHTA new concepts, courses, and initiatives. To provide for continuity, there is no limit on time an individual may serve on the Board. At each AGM, the three directors who have been longest in office since their last appointment must retire but can stand for reelection at the same meeting if they wish. The IOHA representative on the Board is exempt from the retirement requirements.

In 2018, upon review of financial strategy and stakeholder advice, OHTA launched discussions to form a US-based chapter with charity status for the purpose of diversifying its fundraising efforts. The US Chapter Board of Directors is comprised of a chair and four to six directors who are independent of the UK Charity. Its sole role is to raise funds. An overall timeline of the OHTA development process is provided in Table 4.1.

4.4 DELIVERABLES TODAY AT OHTA

4.4.1 COURSES AND MODULES

There are now four levels of courses available:

- Two "Awareness"-level short courses (Silica and Occupational Safety and Health Awareness)

TABLE 4.1

Timeline for OHTA Development

2005	Early informal discussions
	First formal meeting of cross-industry group of senior occupational hygienists to discuss issues
2006	AIOH engaged to assist in identifying candidate to run pilot course for BP. BOHS funded to develop examination for pilot course
	AIOH distributes call for expressions of interest to run a pilot course
	Discussions at BOHS conference
	Pilot course run by the University of Wollongong. Cross-industry group discussion paper circulated widely
	Stakeholder meeting held at AIOH Conference Gold Coast
2007	Presentation to the board of AIHA. The University of Wollongong contracted by BP to author the first two courses.
	Discussions with AIHA Communications and Training Methods Committee
	Presentation to the board of ABIH – formation of the Futures Committee
	Preparation of paper to the IOHA board. Stakeholder meeting at AIHce Philadelphia, on "Global Opportunities for Occupational Hygiene Training and Development."
	Pilot courses run in Australia and Azerbaijan
	Update paper prepared and widely circulated. First of many update "newsletters" circulated
	W502 – Thermal environment completes peer review and is released
2008	W201 – Measurement of hazardous substances completes peer review and is released
	W503 – Noise – piloted
	W504 – Asbestos and Other Fibres – piloted
	W505 – Control of Hazardous Substances – drafted
	Proposal submitted to BOHS conference Bristol inviting BOHS and AIOH to establish legal entity for governance of scheme.
	W506 – Ergonomics Essentials – drafted
	W501 translated into Spanish by BHP and piloted in South America
2009	Further work developing course materials and structure
	Further meetings and discussion, including presentations and national occupational hygiene conferences
	IOHA, AIOH, and BOHS provide funding for development of the website www.OHlearning.com
	Legal formation and registration of OHTA
2010	Launch of OHlearning.com at AIHce and the first Stakeholders Meeting at IOHA conference in Rome, Italy
2011	Brainstorm meeting at BOHS conference Stratford on Avon
2013	Increased effort to translate modules, including Mandarin, Spanish, and French
2014	Service-level agreement signed with BOHS to provide administrative support and the services of a Business Development Manager
	Articles of Association revised
2015	Formal agreement appointing BOHS as an Awarding body and formalizing arrangements for collection of OHTA examination levy
	Sue Davies Scholarship fund established
	Working partnership discussions with Workplace Health Without Borders (WHWB), the American Industrial Hygiene Association (AIHA), the National Institute for Occupational Safety and Health (NIOSH), and the American Society of Safety Professionals (ASSP)
2016	Articles of Association amended. Approval as a Registered Charity by UK Charity Commission
2019	Silica awareness module launched
	AIHA offers online version of W201 Principles course
2020	Online Occupational Safety and Health Awareness course launched.
	US Chapter Nonprofit Company

- Basic "Principles"-level course for people who need a basic understanding of occupational hygiene that also serves as an introduction for those who want to undertake further study.
- Seven "Intermediate"-level modules for technicians, covering practical skills in core aspects of occupational and environmental hygiene.
- Three "Advanced" modules (referred to as specialist courses) to enable transition from technician-level to master's-level study within specific industries. These may also be used in preparation for professional-level accreditation.

The Principles and Intermediate levels use a formative approach to learning which makes assessment an integral part of the learning process. At the Advanced level, assessment may be part of the academic process, but is not formally included in the current certification process. The scheme can thus serve the needs of students from early technician training through to professional development, by a process of spiral learning, where materials are studied to progressively higher levels.

There are currently 13 modules, across the four levels, some of which are translated into a selection of eight languages. The schematics of the course modules are shown in Figure 4.2. Successful completion of six intermediate modules, submittal of a Personal Learning Portfolio, and successfully passing an oral examination can lead to the ICertOH. This complements the IOHA NAR qualifications, offering an intermediate qualification at a lower level focused on the practical aspects of occupational hygiene. ICertOH is a stand-alone qualification and may also provide a stepping-stone for those wishing to progress to NAR certification such as ABIH's Certified Industrial Hygienist (CIH), AIOH's Certified Occupational Hygienist (COH), the Canadian Registration Board of Occupational Hygienists (CRBOH)

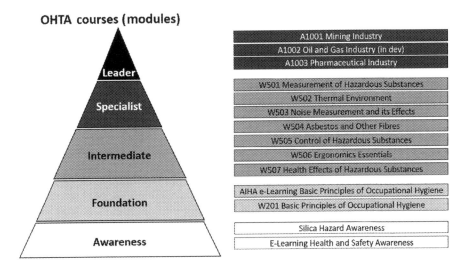

FIGURE 4.2   OHTA course module schematics.

Registered Occupational Hygienist (ROH), and the BOHS Chartered Member of the Faculty of Occupational Hygiene (CMFOH), to name a few.

In addition, OHTA has formed external association and collaborations with WHWB, the ICOH, the WHO, the ILO, the AIHA, the Developing World Outreach Initiative (DWOI), the NIOSH, and the American Society of Safety Professionals (ASSP). These very meaningful and productive relationships have facilitated the development and review of new and existing modules, the development of online modules, and the provision of critically needed occupational and environmental hygiene training in developing economies.

OHTA has also established a foundation for scholarships, and subsequently, the Sue Davies Prize was established to award a prize annually to the best-performing ICertOH student. The prize is financial support for attendance at either the BOHS, AIHA, AIOH, or IOHA conferences.

4.4.2 Distance Learning

One-week modular courses have historically been used for teaching people who are presently working, and the concept is found to be generally acceptable to employers who are more willing to release employees for 1 week at a time rather than for prolonged periods. Traditionally, these have been based on face-to-face teaching and workshop exercises. With rapid advances in technology and more widespread global access to reliable Internet, there is a trend for wider use of technology, with blended learning through videoconferencing, Internet tools, e-learning, and online examinations. Working with partners, the OHTA portfolio now includes an online version of the W201 Principles course developed by AIHA and an online basic Health & Safety Awareness-level course developed by OHTA/Phylmar Academy, both requested by OHTA stakeholders.

This trend to develop distance learning has been accelerated in 2020, with the COVID-19 pandemic. Several OHTA ATPs are now provisionally approved by an OHTA-approved clearinghouse task force to run select distance learning versions of the modules with practical exercises, with BOHS offering online examinations.

4.4.3 Global Outreach – Individual Case Studies

The following illustrate specific case studies that reflect impact by OHTA and its processes:

4.4.3.1 Indonesia

Based in Jakarta, **Indonesia**, Elsye worked in health and safety for a multinational company. With major projects under construction, the business funded Elsye to attend OHTA module training in 2008. With support and mentoring from corporate hygienists, she did such a good job using this knowledge to implement programs in her business that management was willing to fund her to complete an MSc in Occupational Hygiene at the University of Wollongong in 2010. She then earned the ABIH CIH in 2015 and went on in 2016 to found the Indonesian Industrial Hygiene Association (IIHA), becoming its first President with future plans to run OHTA modules.

4.4.3.2   China

The British works manager of a petrochemical complex in **China** had concerns about employee noise exposures. The company was a joint venture between a Chinese company and a Western company. A corporate Industrial Hygiene (IH) based in the region arranged for an early version of the OHTA noise module to be run at the facility. She then established a program of work for the trainees to make use of the skills they had learned in preparing a hearing conservation program, including noise maps and a program of personal dosimetry. With her mentoring, they were able to evaluate the risks and areas requiring action. She also arranged for a noise engineering consultant from the United States to spend time at the site. He designed a program for engineering controls, many based on simple measures. The works manager was so impressed that he willingly supported plans to run other OHTA modules at the site to build up occupational hygiene capability.

4.4.3.3   Former Soviet Union

A new oil field was being developed in a country of the former **Soviet Union** to add to the existing assets, which were all Soviet built. Early in the construction of the offshore rigs and shore facilities, they had experienced health issues and were keen to develop resources to address these. The existing aging Soviet infrastructure in the original platforms and pipelines was also presenting plenty of health and occupational hygiene challenges. Over a number of years, the OHTA Principles course and the full range of OHTA intermediate modules were run on numerous occasions; two occupational hygienists went on to gain the ICertOH, with a third hopefully completing within the next 12 months. Two others are close to applying for the BOHS Diploma to become CMFOHs. In this country, as there are no local regulations for hazardous substances, the company has adopted the UK COSHH Regulations and has developed its own in-house training, based upon the W501 course, to train operators and maintenance technicians to undertake COSHH assessments. Each assessment needs to be authorized by people with more competence and understanding of the issues and, therefore, each COSHH authorizer must pass the W507 Health Effects of Hazardous Substances. This program is ongoing, and with the support of the Central Occupational Hygiene Team, at least one W507 course is run in country every year.

4.4.3.4   Iraq

In **Iraq**, a multinational company had a joint venture with several operators, but the health manager, who is a chartered occupational hygienist working with another contract chartered occupational hygienist, trained up four existing engineers in occupational hygiene and ran the suite of six OHTA modules in Iraq, which resulted in all four gaining the ICertOH qualification.

4.4.3.5   Egypt

In **Egypt**, the local occupational hygienist attended all the OHTA modules with a variety of course providers in the UK. A few years ago, she gained the ICertOH and went on to gaining full BOHS Chartered Status last year. In addition, she won the

BOHS David Hickish Award as the best candidate of that year. She has since run the W201 course to train people involved in a new onshore project, which is building on the W201 courses run in the past, in country, by the companies' corporate occupational hygiene team.

4.4.3.6   Oman

In **Oman**, another big project has led to W201 being run by the UK-based central occupational hygiene team, during which a chemical engineer was identified as a prospective development candidate to become one of the two occupational hygienists needed as backups on the project. This individual has nearly completed all six OHTA modules in the UK by a variety of course providers.

4.4.3.7   South Africa

In **South Africa**, the UK-based central occupational hygiene team of a multinational company has run five W201 courses in Johannesburg to raise awareness of occupational hygiene to people in the downstream oil businesses – safety officers, maintenance engineers, occupational health nurses, permit authorities, terminal managers, etc. People from another of the company's businesses in **Angola** also attended.

4.4.3.8   Russia

In **Russia**, a W201 course was run to train a variety of health and safety practitioners from a joint venture company which partnered with a Western oil company. This was run in Moscow and led by a chartered occupational hygienist from the central occupational hygiene team but supported by two occupational hygienists from the UK North Sea and **Azerbaijan** businesses, respectively. The course was conducted in Russian and required the use of two people providing simultaneous translations. The occupational hygienist from Azerbaijan, who had herself been trained using the OHTA modules, was invaluable on this course as she spoke Russian and was able to help translate the more difficult technical words and concepts. The two translators were also allowed to sit the final examination and were pleased to see that they and everyone else passed. All in all, it was an extremely rewarding if expensive experience, which almost didn't happen because of the difficulties of getting the equipment for the practicals through customs in Moscow airport!

4.4.3.9   Germany

In **Germany**, two W201 courses have been run jointly by a partnership between a UK oil company and a large German agrochemical company. Delegates came from across European businesses, and of the 46 people attending, only one person failed to pass; this was purely down to language difficulties. It is planned to build on this and hopefully run more of the OHTA modules.

4.4.3.10   United Kingdom

In the **UK**, a chemical plant in Hull sent some Health & Safety Executive (HSE) practitioners and laboratory chemists and technicians to join other chemists and analysts from a research center associated with a well-known lubricants business in

Berkshire, for a W501 course. The aim was to develop some competence in monitoring which would allow for a degree of self-sufficiency in both these facilities.

4.4.3.11 Indonesia

In **Indonesia**, there has been a successful history of running OHTA modules, but recently, a multinational company has partnered with occupational hygiene equipment providers, who, in return for supplying the equipment for the practicals, get to have their sales force attend as course participants, thereby increasing their understanding of the equipment they are marketing. This is an initiative the company will be looking to extend into other regions.

4.5 FUTURE VISION

OHTA had been following its long-validated approach to increasing the knowledge and competence level of occupational and industrial hygiene through established ATPs, but new and innovative approaches are now available using technology to provide remote learning. OHTA's first venture at distance training on an e-learning platform was its partnership with the AIHA in late 2019. Although this was an introductory course that does not lead to an international qualification (British/European terminology) or certification (US terminology), this venture was very successful with almost 100 participants in less than the first full year of availability. Meanwhile, the COVID-19 pandemic pushed OHTA forward into distance learning for several of the courses that can lead to the internationally recognized certification (ICertOH).

While dealing with establishing standards for distance learning, OHTA was also approached to develop a short Safety and Health Awareness course that covered the basics of hazard recognition in both health and safety. Two versions of this course have been developed: an e-learning version (no instructor) and a slide version that can be freely downloaded. The e-learning version was developed at cost by a third party (and led by an OHTA Board member), which was another first for OHTA.

The other development has been work on advanced training (beyond the level of intermediate proficiency and at the specialist level). So far, three specialist courses have been developed (Mining; Oil & Gas; and Pharmaceutical industry). Again, this development was based on demand and varies from OHTA's historical approach. Most recently, OHTA entered into a working relationship with several universities to produce additional training funded by US-based government agencies that address some forthcoming new technology issues.

OHTA's future bodes e-learning efforts and more distance learning capabilities and offerings which are very helpful for developing countries that must limit course fees and expensive travel and require local venues for training; more specialty courses at all levels; and more work on emerging health hazards including responses by occupational hygienists to epidemics and pandemics.

OHTA is also looking to enhance its support to students. In countries where there is not widespread familiarity with the principles of occupational and environmental hygiene, there is often a problem recognizing where hazards exist and how best to evaluate all associated risks and develop control strategies that do not give rise to unintended consequences. Knowledge of the scientific principles is not enough. In countries such as the United States, mentoring and support provided role in developing the wider skills required to develop both the science and art of occupational hygiene. By use of case studies and syndicate groups, OHTA tries to help build these skills. However, in the future, OHTA hopes to build on this and examples of students establishing self-help groups by creating the tools to facilitate sharing and mentoring.

OHTA future strategy will need to address these emerging trends. Its plan is to continue working in partnership with stakeholders, such as employers, unions, and others, to identify their needs and develop a wider range of modules and training delivery options. These options will reflect emerging health trends as well as the need for greater use of technology and technological innovation in delivering occupational hygiene training and qualifications.

It is very clear that the work of the OHTA helps to fulfill a significant global need for expanded occupational hygiene competency. The limits of its growth are much more dependent on the ability to raise awareness of global worker illness and injury and the funding support necessary to continue to expand its educational offerings.

REFERENCES

Alesbury, R., Discussion Paper on Industry Needs for Occupational Hygiene (October 2006).

Alesbury, R., Bailey, S., Commentary Addressing the Needs for International Training Qualifications and Career Development in Occupational Hygiene, *Ann Occup Hyg* (2013) 58:140–151.

ILO, 2013; WHO, 2013 and 2016; Armed Conflict, 2016 + Homicide, 2012 (WHO) + Terrorism, 2016 (Statista).

Lucchini, R., London, L., Global Occupational Health: Current Challenges and the Need for Urgent Action, *Ann Global Health* (November 25, 2014) accessed at https://doi.org/10.1016/j.aogh.2014.09.006, October 27, 2020.

Mandela, N., *Oxford Essential Quotations* (5th ed.), edited by Susan Ratcliffe, Oxford University Press (2017), eISBN 9780191843730, accessed at https://www.oxfordreference.com/view/10.1093/acref/9780191843730.001.0001/q-oro-ed5-00007046, October 27, 2020.

WHO, Collaborating Centres in Occupational Health meeting, "Global Strategy on Occupational Health for All: The Way to Health at Work," Beijing, China (11–14, October 1994).

5 The Developing World Outreach Initiative

Expansion of Occupational Hygiene through Volunteerism and Networking

Richard Hirsh
Developing World Outreach Initiative

CONTENTS

5.1   INTRODUCTION

The Developing World Outreach Initiative (DWOI) is an active subcommittee of the American Industrial Hygiene Association (AIHA) Northern California Section (NCS), building capacity in occupational health and safety (OHS) in the developing world for over 15 years. The primary mission of this effort is to address the lack of adequate industrial hygiene resources for workers and health and safety professionals in developing countries by connecting the technical expertise and human resources of the AIHA local section with those who need these resources in the developing world. The target audience for DWOI's activities is defined as nongovernmental organizations (NGOs) – principally OHS associations, universities, and worker and community-based organizations which do not have the resources and connections that government agencies and employers enjoy even in the developing world. The source materials for this chapter rely mostly on the documented DWOI meeting minutes throughout DWOI's history.

The goal of the DWOI committee is to build the capacity of NGOs in developing countries through mobilizing the resources of the local section and other AIHA members to provide information, technical assistance, educational materials, and financial support for small but tangible OHS projects to be conducted by the local NGOs. DWOI continues to create synergies with other like-minded groups such as the Occupational Health Training Association (OHTA, 2020), Workplace Health Without Borders (WHWB, 2020), Occupational Knowledge International (OK International, 2020), and the Maquiladora Health & Safety Support Network (MHSSN, 2020). DWOI also serves as a project team within the AIHA International Affairs Committee (AIHA IAC, 2020), and support of DWOI project requests strengthens the mission of the IAC. Students and young professionals have expressed strong interest in getting involved in global OHS issues, and DWOI projects offer them opportunities to participate. This chapter will provide a history of the origins of DWOI, its initial structure and activities, partnerships, and current efforts.

5.2   ORIGINS OF THE ORGANIZATION

In 1986, directly after graduating from the University of California at Berkeley School of Public Health Environmental Health Sciences program, Richard Hirsh had the opportunity to participate in a volunteer project in Managua, Nicaragua, working on a severe lead exposure situation at a US-owned lead battery manufacturing plant where several workers were diagnosed with acute lead poisoning resulting in hospitalization for much of the workforce. Richard had been active in a group called the Nicaragua Technical Aid Project (NTAP) through the American Public Health Association (APHA, 2020a) which had been contacted by the Ministerio del Trabajo (Ministry of Labor) in Managua, Nicaragua, for assistance on this outbreak of acute lead poisoning cases. The group conducted worker training on the hazards of lead and how to control lead dust levels. This was Richard's first experience involving occupational health in the developing world which greatly influenced his professional trajectory.

Fast-forward to 2005 when Richard was serving as President-Elect of the AIHA-NCS. Since his experience in Nicaragua, he had been employed as a corporate industrial hygienist for the next 20 years and had traveled to nearly 100 chemical manufacturing and research and development sites through many countries in Asia, North, Central and South America, and Europe conducting exposure assessments and audits and providing technical assistance and training to site hygienists. These work trips gave him insights into the challenging working conditions around the world. In his capacity as a local section leader, Richard wondered how the local section could leverage the resources and technical expertise available in the San Francisco bay area and apply them to OHS needs in the developing world. He knew that local section members had textbooks and used equipment, some were even authors of textbooks, some were professors, and some were professionals who traveled globally. The NCS had a diverse membership and money which could be directed to build OHS capacity in the developing world.

Richard decided that the first step in his newly hatched plan was to tap and engage the perspectives of others that had global OHS experiences. Thus, one sunny afternoon Richard invited Garrett Brown and David Zalk to lunch to discuss these ideas. Garrett had been a career Cal/OSHA officer and directed the Maquiladora Health & Safety Support Network (MHSSN), while David had previously served as President of the International Occupational Health Association (IOHA, 2020). Both of these individuals have continued to serve as active advisors to DWOI up to the present.

5.3   INITIAL MEETINGS

The first DWOI meeting was held on March 26, 2006, at the Cal/OSHA offices in the State Building in Oakland, California. Nearly 20 AIHA-NCS local section members attended the first meeting which signaled that there was broad enthusiasm with proceeding on this initiative. At that moment, Richard realized that he had tapped a common calling within his professional colleagues and that there were like-minded volunteers who shared similar goals and were willing to engage in this new endeavor. That was one of the proudest moments of his career. Attendees voiced interests ranging from connecting UC Berkeley students to projects in the developing world to translation of documents, providing technical expertise via the Web, capacity building for OHS personnel, collection of used books and equipment for distribution to NGOs, conducting training workshops, and connecting local section international travelers to universities and other NGOs.

The initial proposal to the Northern California local AIHA section members included the following: "The developing world does not have adequate industrial hygiene resources to address a multitude of health and safety issues facing workers in those countries. There exists a unique opportunity to forge a new relationship with a receptive entity, as yet to be determined, in one or more of those developing countries in order to provide technical industrial hygiene assistance which can be effectively utilized to promote the establishment of industrial hygiene programs where none exist or which are newly established but were in need of additional resources to succeed. Given the leadership and professional connections DWOI have amongst our membership – to AIHA, IOHA, ACGIH, NIOSH, UC Berkeley as well as the wealth of technical expertise and resources in the AIHA-NCS it is our responsibility to set an example for other sections to follow in establishing an outreach mechanism by which DWOI can participate in improving the practice of industrial hygiene (IH) in the developing world" (DWOI, 2006).

The types of assistance that were envisioned consisted of monetary donations, functional IH equipment, IH reference texts, technical exchange services between NCS volunteer members and the chosen recipients, and sponsorship of occupational educational opportunities. Initially, the newly formed committee planned to target government agency OHS departments, university Environmental Health and Safety (EHS) programs, and NGOs directly involved with OHS issues.

DWOI identified potential vehicles for delivery of this assistance though the American Conference of Governmental Industrial Hygienists (ACGIH)-Foundation for Occupational Health and Safety (FOHS) Worldwide Outreach Program (ACGIH FOHS, 2020) and the AIHA IAC. In 2003, the YIHWAG Family Foundation (YIHWAG, 2020) awarded FOHS a substantial grant to benefit the needs of educational organizations in developing countries throughout the world. From this grant, FOHS established the Worldwide Outreach Program. Its mission is to support the professional development of OHS throughout the world. Qualifying educational organizations, professional organizations, and NGOs seeking to further the goals of the World Health Organization (WHO) Collaborating Centers for Occupational Health's 2001–2005 Work Plan could apply for program grants ranging from $1,000 to 2,000. DWOI reviewed the FOHS grant applications that were not funded by FOHS to determine if some projects could be supported through collaboration with DWOI.

By December 2006, DWOI had raised over $2,200, including a $1,000 matching grant, toward funding an as-yet-unfunded FOHS Worldwide Outreach project. Likewise, the AIHA IAC had an International Development Fund and an International Giveaway Fund which could be tapped. Other conduits were also identified within the NIOSH Global Collaborations Program (NIOSH, 2020) and IOHA, the International Committee on Occupational Health (ICOH, 2020), and the International Labour Organization's (ILO) Safe Work Program and Collaboration Centers on Occupational Health (ILO, 2020).

NIOSH had addressed its strategic goal to "enhance global workplace safety and health through international collaborations" by its leadership in the WHO global network of occupational health centers, by partnerships to investigate alternative approaches to workplace illness and injury reduction and provide technical assistance to put solutions in place, by international collaborative research, and by building global professional capacity to address workplace hazards through training, information sharing, and research experience. IOHA conducts a wide range of activities intended to promote and develop occupational hygiene worldwide. From its creation in 1987, IOHA had grown to more than 20 member organizations by the time DWOI was being formed, representing over 200,000 occupational hygienists worldwide. IOHA also cooperates with the work of other international organizations such as ICOH and the International Ergonomics Association (IEA, 2020). IOHA provides an international voice of the occupational hygiene profession through its recognition as an NGO by the ILO and WHO.

One of the key boundaries of the newly formed DWOI was to ensure that it did not draw funds directly from the AIHA local section budget so that the local section could meet its existing financial obligations and to limit its activities to those easily administered with minimal impact to existing AIHA-NCS local programs and activities. It should be noted that since its inception, DWOI has never been given a single budget line item as a cost center within the local section budget. All fundraising has occurred outside the local section membership dues generated annually.

In order to proceed with this initiative, DWOI needed to identify the types of aid it would offer, the list of potential recipients, and the vehicles it would utilize. A formal proposal was drafted and submitted to the AIHA-NCS Board for approval.

5.4   SURVEY OF RESOURCES AND NEEDS

The initial meetings resulted in the development of a survey tool to determine what resources, language and translation capabilities, university connections, mentoring interest, technical skills, international travel destinations, and willingness to participate our internal membership could offer to this effort. This became the responsibility of newly formed DWOI Committee A. Conversely, DWOI wanted to determine what materials (e.g., equipment, reference texts, and technical and training assistance) our external developing world contacts needed. This became the focus of DWOI Committee B.

The Industrial Hygiene Resources Survey sought to capture information on potential volunteers who had expertise or skill sets on training/coaching, editing, document development, translation, and IH experience working in developing countries. Questions were posed on the specific subject matter expertise they offered, language skills, travel plans, affiliations and connections with international organizations and resources, and any equipment, training materials, or reference texts available for donation.

By June 2006, the survey was entered into SurveyMonkey® (SurveyMonkey, 2020) for a test run and critique by DWOI members. The plan was to administer the internal resources survey to the NCS membership over the summer. At the same time, DWOI Committee B was actively casting a wide net toward potential organizations DWOI could partner with and then work with the self-selected respondents to determine areas of needed support. At this time, DWOI also learned about the experiences and activities of several organizations involved in similar efforts to better understand their approaches, successes, projects, and obstacles. This included the MHSSN led by Garrett Brown, the OK International led by Perry Gottesfeld, and the ACGIH FOHS (ACGIH, 2020) and WHO Collaborative Centers presented by David Zalk.

These experiences created significant enthusiasm among the DWOI members as there were several potential avenues to apply our efforts, which were within the capabilities of DWOI which sought to provide tangible, albeit small, contributions to the global efforts already underway. DWOI still needed to determine what types of groups DWOI should work with and how DWOI could leverage our resources effectively and efficiently. DWOI also wanted to decide whether to work with those organizations which already had a proven track record or those groups that lacked such connections or record.

After the survey draft was finalized, it was reviewed and approved by the AIHA-NCS Board. The survey results provided DWOI with a path forward in understanding how DWOI could match and leverage available resources and needs. Forty-six respondents from the local section responded to the internal resources survey. Key potential activities submitted included training/coaching (75%), document development (40%), editing (27.5%), IH experience in specific countries (27.5%), translation (17.5%), and several other ideas. Subject matter expertise was also identified for key IH technical areas including EHS Management Systems, Respiratory Protection, Asbestos, Air Monitoring, Confined Space, H&S Training, and Exposure Assessment Strategies, followed by several other categories. DWOI found that Spanish- and Mandarin-language resources were available among its ranks. Around 20 members could visit with local specialists during international travel.

Volunteers were identified who were willing to deliver presentations/lectures, question-and-answer sessions, training workshops, and document reviews. Several respondents indicated that they had professional affiliation with IOHA, WHO, ICOH, IEC, NIOSH, ACGIH, and local and regional professional EHS organizations, academic/research institutions, and labor unions. Several also indicated that they had reference texts and/or IH equipment to donate. Respondents also stated that they had many available training modules available to share. Thirteen responded that they would be willing to make a donation to DWOI – even though DWOI had not yet formalized what it was DWOI were going to spend it on! There were 52 responses to the final survey which were collected and converted into an Access® database.

5.5   INITIAL COMMITTEE FORMATION

DWOI Committee A, which was focused on internal local section resources, was requested to contact members who had books to donate in order to itemize the list of references available for donation. They were also asked to contact members who had IH equipment to donate and identify the exact equipment, software, and associated user's manuals. The Committee also contacted traveling members to find out their upcoming and routine travel itineraries and schedules. Some members were contacted regarding presentations and educational training materials they would be willing to share in a clearinghouse repository.

DWOI Committee B, which focused on understanding the global network of NGOs and their needs, identified several contacts who provided a list of resource needs and potential NGO contacts. Needs included technical references, especially translated materials, training materials, networking and coalition-building efforts, financial support, guidance on how OHS issues are addressed and controlled in the United States, technical expertise (especially on-site technical assistance), and other capacity-building support. The Asia Monitor Resource Center (AMRC, 2020) and ANROAV (now ANROEV) (ANROEV, 2020) in Hong Kong were identified as potential partners since most DWOI members were traveling to China on business. AMRC would act as the liaison for the ANROAV-affiliated organizations in Asia, who would be asked to alert DWOI of any health and safety projects with which DWOI might be able to assist. ANROEV is an umbrella NGO network of 21 OSH organizations in 14 Asian countries which provide various types of health and safety assistance to local grassroots groups in Asia. DWOI made contact with their principal coordinators, Sanjiv Pandita (Program Coordinator for Occupational Safety and Health at AMRC) and Jagdish Patel (Coordinator at ANROEV). ANROEV along with AMRC has since become a key partner of DWOI to assist with disseminating Requests for Proposals for both technical projects and training workshops.

5.6   DEVELOPMENT OF LIAISONS AND BUSINESS TRAVEL

DWOI also engaged in a dialogue with representatives of both the Center for Occupational and Environmental Health (COEH, 2020) and the University of California at Berkeley Career Center Externship program to inquire about potential collaboration on DWOI projects with graduate students. Although no formal collaboration was created with the centers, DWOI was able to facilitate a project by a UC Berkeley graduate student to conduct a survey of garment workers in the Dominican Republic and facilitate a visit to Honduras and meet with Homero Fuentes, a labor rights leader in Central America. These two student-led projects were conducted by Nina Townsend, a UC Berkeley second-year Master's of Public Health graduate student in the School of Public Health Environmental Health Sciences Industrial Hygiene program. Nina first conducted a survey with the workers from the Gildan Textile Mill (Gildan Textile Mill, 2020) in Guerra, Dominican Republic. Inspired to address the challenges of worker health and safety overseas by DWOI founders, Nina collaborated on a short-term project in the Dominican Republic with the Worker Rights Consortium (WRC, 2020). Nina conducted a survey with workers of the Gildan Activewear Textile Mill to document reports of accidents, fires, and chemical exposures. Workers at the mill were paid piece rate, a number of workers were fired for attempting to organize, and the company refused access to international inspectors. Workers reported a number of hazards including excessively heavy loads, missing machine guards, chemical hazards (including burns), airborne cotton dust, damaged and slippery walking surfaces, and a lack of appropriate personal protective equipment. The findings from the survey were incorporated into a report issued by the WRC and utilized in collective bargaining at a later date.

Another project involved Richard Hirsh's collaboration with Nayati International (Nayati International, 2020) in Hyderabad, India, to serve as an instructor for the OHTA W501 and W505 courses while traveling on business in India. Another DWOI effort involved a member who served to translate the AIHA IH Statistics computer program into Turkish for the AIHA Exposure Assessment Strategies Committee (AIHA EAS, 2020).

Likewise, DWOI member Karen Gunderson, who resettled to Indonesia, connected with the University of Indonesia's Occupational Health and Safety Program (University of Indonesia, 2020) to serve as an instructor in their curriculum. Karen made significant contributions to the field of OHS while living in Jakarta, Indonesia, as an expatriate from 2011 to 2016. Right away, she connected with industrial hygiene and safety professors at the University of Indonesia to give them used industrial hygiene textbooks donated by DWOI members that she had brought in her goods shipment. Then, she arranged for donations of air sampling pumps and calibrators from SKC Inc. (SKC Inc., 2020), an industrial hygiene supplier, which were used by industrial hygiene students in their lab courses. Karen also wrote up sampling instructions, created curriculum, taught IH lab courses, and presented at university safety seminars.

In addition, Karen supported an Indonesian nonprofit grassroots worker safety advocacy group called the Local Initiative for OSH Network (LION, 2020) to build awareness on asbestos hazards in Indonesia. With her assistance, they received grants from DWOI, Korean Green Foundation (2020), and International Ban Asbestos Secretariat (IBAS, 2020); together they conducted asbestos air sampling and hazard assessments at asbestos cement and textile factories. They published a project report on their activities and findings (LION/DWOI, 2014). She also collaborated with them to put on professional asbestos awareness seminars and to produce a video (Dangerous Dusts) that warned the largely ignorant Indonesian population of asbestos hazards. LION's professional work helped them receive a large ongoing grant from the Australian Union Aid Abroad-APHEDA (APHEDA, 2020) program and the establishment of a growing ban asbestos network in Indonesia (INA-BAN, 2020). Even after repatriation, Karen continues to help LION and INA-BAN. Most recently, she edited an Indonesian doctoral student's epidemiology publication showing the high association between asbestos exposure and lung cancer in Indonesian workers (Suraya et al., 2020). Additionally, Karen presented at the American Industrial Hygiene Conference and Exposition (AIHCE) on "Mentored Skill Building Increases Asbestos Hazard Awareness in Indonesia," on June 3, 2015 (Gunderson, 2015).

One DWOI member, Valeria Valasquez of the Labor Occupational Health Program (LOHP, 2020), attended the ANROEV conference in Manila, Philippines, to lead skill-sharing workshops for approximately 150 electronics workers from 13 Asian countries. Workshops focused on chemical hazards, asbestos and silicosis, health effects, reduction and elimination of hazards, safe substitution, and translation of informational fact sheets into local languages. During 2008, DWOI assisted other local section members who traveled to Malaysia, Indonesia, India, and Argentina to coordinate visits with OHS professionals and universities. During 2012, local AIHA-NCS member Mike Cooper visited Cambodia and DWOI connected him with Athit Kong, an organizer and activist in Phnom Penh. They discussed OHS work being done in the garment sector in Cambodia.

5.7 CREATION OF THE DWOI WEBSITE

By September 2006, DWOI had created a webpage within the local section website (www.aiha-ncs.org) which provided a description of the initiative, meeting minutes and agendas for upcoming meetings, a link to the survey, lists of projects, contact information, and plans to include a discussion (Q&A) feature for contacts in the developing world to submit technical questions. Also, at this time, DWOI received an offer from Sandia National Laboratories (2020) in Livermore, California, to donate a large volume of OHS reference texts and training videos. DWOI also reviewed a CD on the topic of silicosis in agate processing and home-based exposures, relationship of silicosis with tuberculosis, and practical controls for silica dust, which was sent by ANROEV.

5.8 PARTNERSHIPS

The partnership plan involved matching traveling AIHA local section members' availability to periodically conduct training or informal seminars with AMRC and

ANROEV. By December 2006, the partnerships with these organizations were realized. AMRC planned to share their existing research and studies in Asia with DWOI, which included information regarding workers' health and participation in OSH decision making and their efforts to work at the grassroots level to make positive changes in the workplace. They also planned to periodically update DWOI regarding ongoing campaigns related to asbestos, coal mining, and electronics hazards. DWOI also planned to leverage these partnerships to identify NGOs under the AMRC and ANROEV umbrellas to host UC Berkeley students in Asia with summer projects coordinated through the UC Berkeley COEH. The partnerships with AMRC and ANROEV were also intended to foster specific activities including but not limited to the following:

1. Provision of written materials on OHS topics of concern to their member organizations.
2. Interaction with the AIHA-NCS local members who travel to Asia who would be available to meet with member organization representatives for mini-workshops or technical assistance and advice.
3. Financial support for small projects.
4. Possible summer visits by students from San Francisco bay area universities to conduct OHS projects with the member organization. This could include risk assessments, industrial hygiene monitoring, training, preparation of verbal presentations, or written hazard information sheets.

The effort to identify external organizations working on global issues with contacts in the developing world resulted in over 40 potential collaborators. Those identified included the African listserv managed by Andrew Cutz in Canada; the AIHA IAC; Agriculture SCOUL (2020) in sub-Saharan Africa; the Directorate of Occupational Health and Safety Services in Kenya (2020); the Tanzania Occupational Health Service (TOHS, 2020); the Department of Pesticides and Toxicology at Gezira University in Sudan, the Occupational Safety and Health Dept Ministry of Gender, Labor and Social Development in Uganda (2020); OK International (San Francisco, CA) (2020); the American Public Health Association Occupational Health and Safety Section (APHA, 2020b); several professors at the University of Washington and University of California at Berkeley which had multiple projects ongoing in China, Vietnam, and India; the Bangladesh Center for Workers Solidarity (BCWS, 2020); the International Ergonomics Association; the ACGIH International Committee (ACGIH, 2020); the Clean Clothes Coalition (Clean Clothes Coalition, 2020); the Maquila Solidarity Network (MSN, 2020); the Network on Climate Change (Network on Climate Change, 2020) in Bangladesh; and other organizations (see Appendix 5.3).

5.9   FORMATION OF SUBCOMMITTEES

By January 2007, after this initial needs assessment and resource inventory was completed, DWOI decided to create four subcommittees: (a) Request and Response, (b) Tools/Resources Website, (c) Liaison, and (d) Publicity and Fundraising. The Request and Response committee focused on books and equipment inventory coordination and management (incoming from section members; shipments to requestors). The Tools/ Resources Website subcommittee focused on building out the DWOI website contents. The Liaison subcommittee focused on outreach to ANROEV/AMRC, the University of California COEH, and DWOI volunteers traveling to developing world countries.

The Publicity/Fundraising effort involved organizing local section dinner meetings involving raffles, silent auctions, and simply "passing the hat" to raise needed funds. The publicity focus involved co-sponsoring local section dinner meetings to update the membership on DWOI efforts, highlight DWOI projects and other global OHS issues via presentations, write articles for the AIHA-NCS local section Monitor newsletter (AIHA-NCS, 2016) and national AIHA Synergist (AIHA, 2020), and speak at various conferences and other forums. DWOI current members agreed to request that all members on the distribution list recommit their participation on at least one of the four newly formed subcommittees. DWOI also finalized our decision-making process and finalized our speakers for an upcoming March dinner meeting.

By February 2007, DWOI had solicited project proposals from Thailand on control banding, India on an occupational hygiene library, Cameroon on safety issues, Hong Kong on participatory training, Nigeria on urinary lead testing, and Malaysia on an Indoor Air Quality course. These projects were reviewed and scored for merit and feasibility, and final recommendations for funding were completed by March 2007. DWOI provided $2,000 for the Malaysian Industrial Hygiene Association (MIHA, 2020) channeled through the ACGIH FOHS. The grant was used for purchasing and distributing technical reference materials on Indoor Air Quality (IAQ) issues. These materials were distributed at their local IAQ course to bolster the IH profession in the region, regarding this technical area, and increase the number of participants with this important reference tool at no additional charge to attendees.

This effort served as an excellent example of promoting sustainability and self-sufficiency for our profession in the region. In addition, the MIHA is a member of IOHA and these funds assisted the MIHA in gaining greater reach and respectability among the OHS science practitioners. The remaining $600 raised during the fund drive was used to cover shipping costs related to textbook shipments to the University of Calabar in Nigeria (2020), the TOHS in Tanzania, and the University of Gezira in Sudan. In addition, a connection was made with SKC Inc. to provide sampling pumps to Dr. Bassey Offiong Ekpo at the Department of Pure and Applied Chemistry at the University of Calabar in Nigeria through their equipment donation program.

Also, during this time frame, a dialogue began between DWOI and professors at the UC Berkeley School of Public Health as well as the coordinator for the COEH based at UC Berkeley to identify student-led projects and build a closer relationship to coordinate COEH-sponsored student awards through participation in overseas projects. Besides that effort, DWOI also established additional goals for 2007 which included writing an AIHA Synergist article, setting fundraising targets, attending the ANROEV conference, identifying projects in Central and South America, targeting several textbook shipments, a presentation to the AIHA IAC, promoting the concept of DWOI to other local AIHA sections, developing a procedure for website postings, and sponsoring NGO participants from China at an AIHA event. By this time, DWOI membership stood at 25 members.

The DWOI website procedure, developed in April 2007, allowed Web-based postings of various industrial hygiene resources including tools, organizational links, and informational and training materials as well as meeting agendas and minutes. The targeted audiences were AIHA-NCS members who provide industrial hygiene services to workplaces with limited financial resources and/or few English-speaking workers. The second audience involved industrial hygienists, students, and NGOs working in developing countries. The process would be managed by the AIHA-NCS webmaster, the DWOI website chairperson, and website subcommittee.

By July 2007, DWOI became a Volunteer Group Project Team of the AIHA IAC. The project team supported DWOI's efforts to provide information and technical assistance to NGOs in the developing world. Since DWOI had established ongoing contacts with NGOs in both Asia and Africa who were interested in English-language written and cyber materials, it seemed a logical extension of DWOI's mission to collaborate through the AIHA IAC.

The goal of the project was to provide NGOs who were concerned about and active on workplace OHS issues with information, training materials, and technical assistance to increase their capacity to understand OHS issues, to design and implement effective interventions to improve workplace conditions, and to increase the technical quality of their educational efforts. The project purpose was to build the profession in developing countries where existing expertise was limited or nonexistent, and to do so in a manner that would be self-sustaining and broad-based.

Initial specific deliverables of the IAC DWOI project team were to (a) send three to five shipments of written reference materials and equipment to occupational health programs at universities in Africa and Asia; (b) expand the recently established DWOI webpage on the NCS Section website to include more links to OHS websites and electronic resources in various languages; (c) to post model safety programs, lecture outlines, and training curriculum; and (d) to post conference presentations made by local section members on various OHS subjects.

Also, in July 2007, Roger Alsbury, the then Director of Industrial Hygiene at British Petroleum, wrote to Garrett Brown at DWOI regarding an effort underway to develop IH training materials for use in those countries with little IH resources or infrastructure. Roger was excited to see the DWOI report to the AIHA IAC on the progress made to engage groups internationally. He had been collaborating with other colleagues in multinational corporations to share information on how best to develop IH capability through improvements in knowledge, understanding, and application of industrial hygiene. British Petroleum was sponsoring the development of a number of modular training courses that would be made widely available. The plan was to engage stakeholders including AIHA and IOHA members to help support this effort. Given the common vision, Roger requested DWOI to consider posting the training materials, once available, on the DWOI website. Roger also shared the minutes of the AIHCE meeting from June 4, 2007, on Global Opportunities for Occupational Hygiene Training and Development. As DWOI now know, this commendable effort along with several training modules eventually became available by the OHTA which Roger Alsbury and Steven Bailey of GlaxoSmithKline subsequently formed.

5.10   DINNER MEETING SPEAKERS/TOPICS/FUNDRAISERS

At least once per year, DWOI has co-sponsored an AIHA Northern California local section dinner meeting. The first co-sponsored dinner meeting occurred in March 2007 at the now-closed Spenger's Fish Grotto in Berkeley, California, which coincided with its annual "Student's Night" in which the local section recognized two outstanding students from nearby universities with the Tebbens and Legge Awards. Natalia Varshavsky opened the dinner meeting with an overview of the newly formed DWOI and its planned activities. Garrett Brown volunteered to speak on the new AMRC/ANROAV partnerships, Pam Tau Lee volunteered as a guest speaker on participatory training programs and experiences, and Brian Daly, the then chair of the AIHA IAC, volunteered as the guest speaker to outline the committee's international work effort. For another dinner meeting held in November 2009, Miriam Lara-Meloy from the Hesperian Foundation described their foundations published "Workers' Guide to Health and Safety" (Jailer, Lara-Meloy, and Robbins, 2009). Amanda Hawes from the Center for Occupational Safety and Health (COSH, 2020) discussed the precautionary principle at work by examining historical events. In March 2012, DWOI co-sponsored an AIHA-NCS dinner meeting highlighting current IH challenges in Mexico and the role of the Mexican Industrial Hygiene Association (AMHI, 2020) in addressing these challenges.

These annual dinner meeting events serve to highlight current DWOI activities as well as showcase a guest speaker whose topic is on an international OHS theme. The dinner meetings also serve as fundraisers to support training workshops, technical projects, and AIHA international affiliate memberships. Previous fundraisers also helped support our reference book shipment program until it became too resource-intensive to continue.

5.11   AIHA INTERNATIONAL AFFILIATE MEMBERSHIPS

One of the focus areas involves supporting our partner NGO organizations with AIHA international affiliate memberships which provides access to several AIHA offerings including reference texts, webinars, the Synergist® magazine, members-only content, online community, and resources. International members may vote on association-related issues and may serve on committees. It provides these NGOs with a mechanism to keep current on global reaching topics, technology, and trends as well as access to a network of global Occupational and Environmental Health and Safety (OEHS) practitioners to assist with identifying solutions to issues. Typically, DWOI has been able to sponsor three to five international affiliate memberships each year for representatives of the AMRC (Omana George, Sanjiv Pandita), ANROEV (Noel Correa), and other organizations.

5.12   DISTRIBUTION OF REFERENCE TEXTS

DWOI learned through our original survey that several local members had bookshelves full of OHS reference textbooks, some having sat on shelves for years as newer editions and e-books became available. Thus began the effort to collect books and distribute them to fledgling university libraries and other NGOs on a first-come-first-serve basis. By June 2007, DWOI had collected several hundred key reference books and other teaching materials (including industrial hygiene equipment) which were cataloged for redistribution. The primary targets for distribution of the reference texts were universities in Africa which was chosen because the continent was not a focus area of AIHA international outreach efforts at the time.

DWOI also raised funds to ship these materials. They were able to collect and store the books at the Oakland California Cal/OSHA offices, and DWOI volunteers boxed and cataloged the texts. They then sent the current list out to the DWOI contact list and potential recipients who then submitted their requests which were filled as feasible depending on resources, on a first-come-first-serve basis. Over several years, DWOI found that this effort was quite labor-intensive and costly. Between 2007 and 2017, they shipped thousands of books around the globe, primarily to Africa and Asia including the following countries: Afghanistan, Bhutan, Botswana, Cameroon, Chile, Ethiopia, Ghana, Grenada, India, Indonesia, Kenya, Malaysia, Namibia, Nicaragua, Nigeria, Pakistan, Singapore, South Africa, Sudan, Tanzania, Turkey, Uganda, and Zambia.

By October 2008, DWOI was collaborating with the AIHA Social Concerns Committee (AIHA SCC, 2020) and the IAC who were interested in developing a program to ship books to developing countries via a central hub at an African university and the AIHA Board was willing to fund the project. The plan involved establishing partnerships with NGOs which focus on donation of scientific books and resources to international communities and solicit corporate participation in securing new resources. Unfortunately, this AIHA project was never fully realized or initiated due to a lack of logistical coordination and volunteers to orchestrate the endeavor.

Due to the high cost of the actual shipments, custom duties, and unforeseen mishaps such as shipments getting lost in transit or not picked up at the intended destination, DWOI explored more economical methods of transferring OHS knowledge. Our last physical shipment of books, in 2017, to the University of Pakistan in Lahore, involved about ten boxes containing over 400 books at the request of Dr. Mohammed Akram who was serving to establish an OHS program there as a professor. That shipment took about 4 months to complete and involved difficult negotiations with shipment companies and middlemen who wanted payments along the way. Thus began our collaboration with AIHA National to offer e-books of select references via a Google account passcode whereby recipients could download Adobe Acrobat® (.pdf) versions free of charge. Currently, about 60 reference texts have been made available by AIHA through the DWOI portal. At the time of this writing, DWOI is rejuvenating this effort and currently discussing a similar collaboration with ACGIH.

5.13   DISTRIBUTION OF INDUSTRIAL HYGIENE EQUIPMENT

DWOI has also been involved with coordinating donations of industrial hygiene equipment. Initially, DWOI accepted donated used equipment and transferred it to requesting NGOs and universities; however, DWOI faced the reality of costly customs duties, extensive shipment paperwork, and equipment repair issues. In some cases, equipment accompanied DWOI members who were traveling to recipient destinations to avoid these obstacles. One key partnership has been with SKC Inc.

which has an SKC Industrial Hygiene Degree Program Equipment Grant program whereby universities can apply for IH equipment and sampling media. DWOI has collaborated with SKC to provide new IH equipment to multiple universities in Asia, the Caribbean, and Africa through this program. Equipment donations (both SKC University Equipment Grant Program and used equipment) have been sent to university programs in Nigeria, Indonesia, Grenada, and South Africa. Recently, International Safety Systems Inc. (ISS, 2020) loaned equipment for a silica exposure assessment project in India and the Belgian Center for Occupational Hygiene (BECOH, 2020) loaned sampling equipment for a silica exposure project in Tanzania.

5.14 DONATED LABORATORY ANALYSIS

DWOI has had some success with AIHA-accredited laboratories to donate sampling media and sample analysis of samples submitted on behalf of NGOs and universities conducting technical projects. In the past, free analysis was provided for an asbestos survey in Indonesia and silica exposure assessments in both India and Tanzania. Most recently, the BECOH, SGS Galson (SGS, 2020), and Forensic Analytical Laboratories (Forensic Analytical, 2020) donated services to analyze samples for asbestos, volatile organic solvents, diesel exhaust, and welding fume.

5.15 TRAINING WORKSHOPS AND TECHNICAL PROJECTS

Initially, DWOI collaborated with a consulting firm, Golder Associates (Golder Associates, 2020), which offered DWOI discounted "two for the price of one" training workshops in Asia so that DWOI could sponsor representatives from NGOs to attend. These workshops occurred in Guangzhou, Shanghai, Qingdao, and Shenzhen, China, as well as in Mumbai and Pune, India; Singapore; Jakarta, Indonesia; and Manila, Philippines. Although these were successful events, DWOI could only sponsor one or two persons to attend a given training. Thus, over the years DWOI transitioned our efforts toward sponsorships of group trainings for multiple workers or worker representatives conducted by a local NGO, once the NGO submitted to us a training curriculum that was vetted and approved by DWOI. This allowed us to utilize our limited funding to reach a much larger training audience. Similarly, DWOI began requesting submissions for technical projects involving exposure assessments and worker surveys to sponsor using the same vetting and approval process.

Over the last 10 years, DWOI has collaborated on several training workshops and technical projects, typically sponsoring 5–15 trainings and 3–7 technical projects each year depending on funding. Some of the collaborating NGOs include ANROEV in Hong Kong, the Metal Workers Alliance of the Philippines (MWAP, 2020), Cividep Workers' Rights and Corporate Accountability (CIVIDEP, 2020) in India, Mallarpur Uthnau in West Bengal, the University of Nairobi (University of Nairobi, 2020) in Kenya, Burhanpur Powerloom Bunkar Sangh (BPBS) in India, and Jeevan Rekha Parishad (JRP, 2020) in India, among others. (See the list of NGO contacts for training and technical grant requests for proposals in Appendix 5.3.) Select project topics included cotton dust among loom weavers in India, asbestos exposure in Indonesia, metal working hazards in the Philippines, and promoting the occupational hygiene profession in Uruguay.

FIGURE 5.1 Gold mercury amalgamation during artisanal mining in Tanzania. (Photo courtesy of Jeff Dalhoff.)

Other projects, including stonecutter silica exposure in India, lead in battery recycling in Kenya, and silica exposure during artisanal gold mining in Tanzania, involved collaborations with OK International. The project on silica exposure during gold mining in Tanzania resulted in both a published *Journal of Occupational and Environmental Hygiene* article entitled "Silica Exposures in Artisanal Small-Scale Gold Mining in Tanzania and Implications for Tuberculosis Prevention" by Gottesfeld, Andrew, and Dalhoff (2015) and a presentation at the AIHCE in Salt Lake City in June 2015 entitled "Silica Exposures in Artisanal Small-Scale Gold Mining in Tanzania" (Dalhoff, 2015). Figure 5.1 shows gold mercury amalgamation processes during artisanal gold mining in Tanzania.

5.16 PUBLICITY

As early as October 2007, AIHA editorial staff indicated that a submitted DWOI article entitled "Developing and Retaining IH Talent Globally" (Hirsh, 2007) would be published as a short item in the January 2008 Synergist and the entire article would be published in the AIHA Leadership Newsletter in December 2007 (Hirsh, 2007). This material would also be presented at the AIHCE, Minneapolis, MN, in 2008 (Hirsh, 2008).

DWOI has continued to gain publicity through AIHA Synergist articles, presentations at the AIHCE and California Industrial Hygiene Council (CIHC) Professional Development Seminar (Hirsh, 2008), and the Phylmar Regulatory Roundtable (Hirsh, 2014). In May 2009, Richard Hirsh presented "Global Supply Chains and Corporate Responsibility – Efforts to Develop Global IH Expertise from within an AIHA Local Section" (Hirsh, 2009) on DWOI activities at the AIHCE in Toronto, Canada. In 2011, DWOI had the article "The AIHA-NCS Developing World Outreach Initiative – A Case Study in Technology Transfer Activities" published in the African Newsletter on Occupational Health and Safety (Hirsh, 2011). AIHA published the article in the Synergist under the title "A Model Initiative – How an AIHA Local Section Spreads

OHS Knowledge Worldwide" in November 2011 (Hirsh, 2011). In the June/July 2012 Synergist edition, an Honor Roll article entitled "DWOI Makes Global Impact on OHS" highlighted DWOI activities globally (Bechtold, 2012). Additionally, an article entitled "AIHA Honors International Outreach Initiative" (COEH, 2013) was published by the University of California COEH (Berkeley, Davis and San Francisco) in their Winter 2013 edition of Bridges which highlighted the 2012 AIHA Social Responsibility Award that DWOI received that year (AIHCE, 2012).

5.17   CURRENT TECHNICAL AND TRAINING PROJECTS

DWOI continues to sponsor NGOs and universities in the developing world to conduct technical projects and training workshops. Typically, DWOI receives 20–30 project proposals requesting funding for technical grants and an equivalent number for training workshops. Lists of various grants received in the past several years are shown in Appendices 5.1 and 5.2. With financial support from AIHA, MHSSN, and WHWB-US, as well as local section members and other generous donors, DWOI has been able to fund 3–5 technical projects and 10–15 training workshops in recent years with around $8,000–$10,000 per year. As in the past, DWOI shares project outcomes through the AIHA NCS website, through published papers in the *Journal of Occupational and Environmental Health* (JOEH), presentations at AIHCEs, articles in the AIHA Synergist, and other venues and forums. To date, DWOI has reached 27 developing countries with our support efforts.

During 2018–2019, DWOI issued 3 Technical Grants and 14 Training Grants. During this current funding cycle (2019–2020), DWOI issued 5 Technical Grants and 12 Training Grants. DWOI have built relationships with MHSSN and WHWB International and the WHWB-US to collaboratively funded projects. The AIHA Microgrants Committee provided $3,258 to fund two training workshops and one technical project as well as an additional $5,000 toward other DWOI 2019–2020 projects and $7,500 for 2020–2021 future projects. Requests for Proposals will be issued in September 2020, and projects will be funded in 2021.

DWOI has sponsored over 65 training grants since 2008 in the following countries: Bangladesh, Brazil, Cambodia, China, Hong Kong, India, Indonesia, Kenya, Lebanon, Mexico, Mozambique, Nepal, Philippines, Thailand, Singapore, Sri Lanka, and Vietnam.

Recent training workshops include the following sponsorships:

- LION in Indonesia: "Workplace Hazards – Women Workers – OHS Principles, reproductive health, body/hazard mapping."
- Worker's Initiative in Kolkata, India: "Industrial Worker Training covering several industry sectors focused on electrodes for welding, electrical hazards, chemical hazards and asbestos."
- Comité Fronterizo de Obrer@s – Border Committee of Workers in Mexico: "The Reality of OHS in 4 Mexican Maquiladora Factories in 2018. How Appropriate They Are?"
- CIVIDEP in India: "Hazard Mapping Training Program for Garment Workers on Occupational Health and Safety in Bangalore, Karnataka."

- Center for Development and Integration in Vietnam: "Training on mapping and identifying OHS risks at workplaces in the electronics industry in Bac Ninh Province."
- Workers' Assistance Center in the Philippines: "Basic Workshop and Training on OSH in Cavite EPZs Companies (focused on electronics and semi-conductor industry workers): A Basic Training of Trainers."

Examples of recent small technical project grants that DWOI has sponsored included the following:

India:
- Silica Exposure amongst Stone Crushers;
- Byssinosis and Tuberculosis amongst Power Loom Weavers;
- Tea plantation hazards;
- Mapping home-based worker hazards;
- Environmental monitoring for an asbestos plant.

Indonesia: Asbestos Exposure during Manufacturing; Electronics worker biomonitoring

Tanzania: Silica Exposure in Gold Mining Operations and Silica Exposure amongst Stone Crushers (with WHWB and OK International)

Kenya: Lead Exposure during Battery Recycling; Artisanal gold miners and mercury hazard reduction efforts (with OK International)

Lebanon: American University of Beirut exposure assessments

Nepal: Mercury exposure assessment

Nigeria: Mercury and Hydroquinone Exposures in Beauty Parlors

South Africa: Acrylate and VOC Exposures in Nail Salons

5.17.1 PROJECT REVIEW AND RATING PROCESS

Requests for Project Proposals are sent to NGOs and universities in early September each year for both Technical Project and Training Workshop grants. The project proposals are then submitted by NGOs and universities by November 1 of that year. A DWOI volunteer review committee (six to eight reviewers) is formed and reviews and rates each project independently and then tabulates ratings and reaches consensus on which projects to fund based on available resources. Funds are distributed to recipients in December for projects beginning in January of the following year.

5.17.2 PROJECT PROPOSAL REVIEW CRITERIA

DWOI has recently revised its review criteria to ensure transparency and clarity for the review process. The eligibility of each organization (e.g., NGO, university) is considered first. DWOI then evaluate whether the submittal is a stand-alone project versus a segment of a larger project and determine whether it is a technical versus training project as some submittals have truly been training workshops submitted as technical projects. For training workshops, DWOI review the proposed curricula. DWOI also evaluate whether the budget items are well defined and justified and total less than our project limit of $2,000 per technical project and $500 per training workshop. Allowable expenses include, but are not necessarily limited to, the following:

- Equipment, supplies, and media for sample collection
- Sample-shipping and analytical costs (e.g., to laboratory for chemical analysis)
- (Local) travel to site of project execution
- Participant fees (token for cooperation beyond normal work duties) if applicable
- Modest salary or stipend specifically for preparation or execution of project tasks (note that salaries and stipends must be <50% of the budget).

For technical projects, the next criterion is to evaluate whether the project aims are focused on workplace exposure characterization, prevention, control, and/or capacity building related to recognized hazards. The project must clearly describe the work plan with sufficient scope and merit as well as the feasibility of meeting specific aims in a specified time frame (including adequacy of resources). DWOI considers the nature and extent of the impact on worker populations and/or local/regional occupational hygiene capacity. Finally, DWOI considers the qualifications/preparation of the key investigator(s) and/or past achievements of the organization and may offer inclusion of technical assistance or collaboration with other local organizations or DWOI based on our internal expertise.

5.18 THE FUTURE OF DWOI

Despite operating on barely a shoestring of funding, DWOI remains active and focused on current projects during the 2020–2021 funding year. After 15 years, DWOI remains a committed and effective group of dedicated professionals focused on carrying out DWOI's stated mission to build capacity on OHS in the developing world. DWOI will continue to collaborate with like-minded organizations such as MHSSN, OK International, and WHWB on these efforts. Most recently, DWOI has collaborated with Professor Kathy Hammond at the UC Berkeley School of Public Health to create the first-ever Global Occupational Health and Safety Course (PH 290), which was held in the fall semester of 2018 and again in the spring semester of 2020 enrolled with 10–12 students. DWOI members Garrett Brown, Nina Townsend, Perry Gottesfeld, Jeff Dalhoff, Karen Gunderson, and Richard Hirsh, among others, participated in creating and delivering the curriculum. Current discussions are ongoing with University of Michigan's Professor Ted Zellers (also a DWOI member) and other professors on how to share lecturers as well as lectures between various OHS graduate programs. DWOI continues to think globally and act locally!

REFERENCES

ACGIH, American Conference of Governmental Industrial Hygienists (2020) acquired at https://www.acgih.org/ accessed on May 1, 2020.

ACGIH-FOHS, Foundation for Occupational Health and Safety, Worldwide Outreach Fund (2020) acquired at https://www.acgih.org/foundation/programs/worldwide-outreach-program accessed on May 1, 2020.

Agriculture SCOUL (2020) acquired at https://en.wikipedia.org/wiki/Sugar_Corporation_of_Uganda_Limited accessed on May 1, 2020.

AIHA, Synergist (2020) acquired at https://www.aiha.org/publications/the-synergist accessed on May 1, 2020.

AIHA EAS, Exposure Assessment Strategies Committee IH Statistics Application, IH Apps & Tools (2020) acquired at https://www.aiha.org/public-resources/consumer-resources/topics-of-interest/ih-apps-tools accessed on May 1, 2020.

AIHA IAC, AIHA Social Responsibility Award (2012), AIHCE, Montreal.

AIHA IAC, International Affairs Committee, Volunteer Groups (2020) acquired at https://www.aiha.org/get-involved/volunteer-groups accessed on May 1, 2020.

AIHA-NCS, Local Section Monitor Newsletter (2016) acquired at https://aiha-ncs.org/images/news/AIHA-NCS_Newsletter_April_2016.pdf accessed on April 5, 2021.

AIHA SCC, Social Concerns Committee, Volunteer Groups (2020) acquired at https://www.aiha.org/get-involved/volunteer-groups accessed on May 1, 2020.

AMHI, Mexican Industrial Hygiene Association (2020) acquired at https://amhi.org.mx/ accessed on May 1, 2020.

AMRC, Asia Monitor Resource Center (2020) acquired at https://www.amrc.org.hk/ accessed on May 1, 2020.

ANROEV, Asia Network for the Rights of Occupational Accident Victims (ANROAV); now called the Asia Network for the Rights of Occupational and Environmental Victims (ANROEV) (2020) acquired at http://www.anroev.org/ accessed on May 1, 2020.

APHA, American Public Health Association (2020a) acquired at www.apha.org accessed on May 1, 2020.

APHA, American Public Health Association, Occupational Health and Safety Section (2020b) acquired at https://www.apha.org/apha-communities/member-sections/occupational-health-and-safety accessed on May 1, 2020.

APHEDA, Australian Union Aid Abroad-APHEDA (2020) acquired at https://www.apheda.org.au/ accessed on May 1, 2020.

BCWS, Bangladesh Center for Workers Solidarity (2020) acquired at https://www.solidaritycenter.org/tag/bangladesh/ accessed on May 1, 2020.

Bechtold, K., "DWOI Makes Global Impact on OHS" (June/July 2012), Honor Roll edition, AIHA Synergist.

BECOH, Belgian Center for Occupational Hygiene (2020) acquired at https://www.becoh.be/nl/ accessed on May 1, 2020.

CIVIDEP Workers' Rights and Corporate Accountability (2020) acquired at http://www.cividep.org/ accessed on May 1, 2020.

Clean Clothes Coalition (2020) acquired at https://cleanclothes.org/ accessed on May 1, 2020.

COEH, Center for Occupational and Environmental Health, "AIHA Honors International Outreach Initiative" (2013), University of California Center for Occupational and Environmental Health, Berkeley, Davis and San Francisco, COEH Bridges Newsletter, Winter 2013 edition.

COEH, Center for Occupational and Environmental Health (2020) acquired at https://www.coeh.berkeley.edu/ accessed on May 1, 2020.

COSH, Center for Occupational Safety and Health, Local COSH Groups (2020) acquired at https://www.coshnetwork.org/COSHGroupsList accessed on May 1, 2020.

Dalhoff, J., "Silica Exposures in Artisanal Small-Scale Gold Mining in Tanzania (June 2015), AIHCE, Salt Lake City, UT, accessed on May 1, 2020.

Directorate of Occupational Health and Safety Services in Kenya (2020) acquired at https://labour.go.ke/directorate-of-occupational-safety-and-health-services-doshs/ accessed on May 1, 2020.

Forensic Analytical Laboratories (2020) acquired at http://www.falaboratories.com/ accessed on May 1, 2020.

Gildan Textile Mill in Guerra, Dominican Republic (2020) acquired at https://panjiva.com/Gildan-Activewear-Dom-Rep-Textile-C/1468907 accessed on May 1, 2020.

Golder Associates (2020) acquired at https://www.golder.com/ accessed on May 1, 2020.

Gottesfeld, P., Andrew, D., Dalhoff, J., "Silica Exposures in Artisanal Small-Scale Gold Mining in Tanzania and Implications for Tuberculosis Prevention" *Journal of Occupational and Environmental Hygiene*, Volume 12, 647–653, 2015.

Gunderson, K., "Mentored Skill Building Increases Asbestos Hazard Awareness in Indonesia" (June 3, 2015), AIHCE.

Hirsh, R., "Developing and Retaining IH Talent Globally" (December 2007), AIHA Leadership Newsletter.

Hirsh, R., "Developing and Retaining IH Talent Globally" (2008), AIHA Synergist.

Hirsh, R., "Developing and Retaining IH Talent Globally" (2008), AIHCE, Minneapolis, MN.

Hirsh, R., "Developing and Retaining IH Talent Globally" (2008), California Industrial Hygiene Council (CIHC) Professional Development Seminar.

Hirsh, R., "Global Supply Chains and Corporate Responsibility – Efforts to Develop Global IH Expertise from within an AIHA Local Section" (2009), AIHCE, Toronto, Canada.

Hirsh, R., "A Model Initiative – How an AIHA Local Section Spreads OHS Knowledge Worldwide" (November 2011), AIHA Synergist.

Hirsh, R., "The AIHA-NCS Developing World Outreach Initiative – A Case Study in Technology Transfer Activities" (2011), *African Newsletter on Occupational Health and Safety*, Volume 21, Number 2, August 2011.

Hirsh, R., "Developing and Retaining IH Talent Globally" (2014), Phylmar Regulatory Roundtable.

IBAS, International Ban Asbestos Secretariat (IBAS) (2020) acquired at http://www.ibasecretariat.org/ accessed on May 1, 2020.

ICOH, International Committee on Occupational Health (2020) acquired at www.icoh.org and http://www.icohweb.org/site/homepage.asp accessed on May 1, 2020.

IEA, International Ergonomics Association (2020) acquired at https://iea.cc/ accessed on May 1, 2020.

ILO, International Labour Organization's Safe Work Program and Collaboration Centers on Occupational Health (2020) acquired at https://www.ilo.org/safework/cis/lang--en/index.htm accessed on May 1, 2020.

INA-BAN (2020) acquired at https://inaban.org/ accessed on May 1, 2020.

IOHA, International Occupational Health Association (2020) acquired at https://www.ioha.net/ accessed on May 1, 2020.

ISS Inc., International Safety Systems Inc. (2020) acquired at https://issehs.com/ accessed on May 1, 2020.

Jailer, T., Lara-Meloy, M., Robbins, M., *Workers' Guide to Health and Safety* (2015), Hesperian Foundation, acquired at https://store.hesperian.org/prod/Workers_Guide_to_Health_and_Safety.html accessed on May 1, 2020.

JRP, Jeevan Rekha Parishad (2020) acquired at https://www.jrpsai.org/ accessed on May 1, 2020.

Korean Green Foundation (2020) acquired at http://www.greenfund.org/en/m12.php accessed on May 1, 2020.

LION, Local Initiative for OSH Network (2020) acquired at www.lionindonesia.org accessed on May 1, 2020.

LION/DWOI, "Factory Exposure Assessment, PT. Siam Indo Karawang Indonesia, Process and Activity Report" (2014), Local Initiative for OHS Network (LION) Indonesia and the Developing World Outreach Initiative.

LOHP, University of California Labor Occupational Health Program (2020) acquired at https://lohp.org/ accessed on May 1, 2020.

MHSSN, Maquiladora Health and Safety Support Network (2020) acquired at http://mhssn.igc.org/ accessed on May 1, 2020.

MIHA, Malaysian Industrial Hygiene Association (2020) acquired at https://miha2u.org/ accessed on May 1, 2020.

MSN, Maquila Solidarity Network (2020) acquired at https://www.maquilasolidarity.org/ accessed on May 1, 2020.

MWAP, Metal Workers Alliance of the Philippines (2020) acquired at http://philmetalworkers.org/ accessed on May 1, 2020.

Nayati International (2020) acquired at http://www.nayati.org/ accessed on May 1, 2020.

Network on Climate Change (2020) acquired at http://www.climatenetwork.org/ accessed on May 1, 2020.

NIOSH, National Institute of Occupational Health and Safety (2020) acquired at www.cdc.gov/niosh accessed on May 1, 2020.

NIOSH, National Institute of Occupational Health and Safety, Global Collaborations Program (2020) acquired at https://www.cdc.gov/niosh/docs/2016-138/default.html accessed on May 1, 2020.

Occupational Safety and Health Dept Ministry of Gender, Labor and Social Development in Uganda (2020) acquired at https://www.facebook.com/mglsd/posts/ministry-of-gender-labour-and-social-development-department-of-occupational-safe/557856531024264/ accessed on May 1, 2020.

OHTA, Occupational Health Training Association (2020) acquired at http://www.ohlearning.com/about-ohta/purpose-and-principles.aspx accessed on May 1, 2020.

OK International, Occupational Knowledge International (2020) acquired at www.okinternational.org accessed on May 1, 2020.

Sandia National Laboratories (2020) acquired at https://www.sandia.gov/ accessed on May 1, 2020.

SGS Galson (2020) acquired at https://www.sgsgalson.com/ accessed on May 1, 2020.

SKC Inc., Industrial Hygiene Degree Program Equipment Grant Application (2020) acquired at www.skcinc.com accessed on May 1, 2020.

Suraya, A., Novak, D., Widajati-Sulistumo, A., Ghanie Icksan, A., Syahruddin, E., Berger, U., Bose-O'Reilly, S., 2020, "Asbestos-Related Lung Cancer: A Hospital-Based Case-Control Study in Indonesia", *International Journal of Environmental Research and Public Health*, Volume 17, 591, 2020, doi:10.3390/ijerph17020591 acquired at www.mdpi.com/journal/ijerph.

SurveyMonkey® (2020) acquired at www.surveymonkey.com accessed on May 1, 2020.

TOHS, Tanzania Occupational Health Service (2020) acquired at http://www.tohs.or.tz/ accessed on May 1, 2020.

University of Calabar in Nigeria (2020) acquired at https://www.unical.edu.ng/ accessed on May 1, 2020.

University of Indonesia, Occupational Health and Safety Program (2020) acquired at https://www.ui.ac.id/en/master/occupational-health-and-safety.html accessed on May 1, 2020.

University of Nairobi in Kenya (2020) acquired at https://www.uonbi.ac.ke/ accessed on May 1, 2020.

WHWB, Workplace Health Without Borders (2020) acquired at www.whwb.org accessed on May 1, 2020.

WRC, Worker Rights Consortium (2020) acquired at https://www.workersrights.org/ accessed on May 1, 2020.

YIHWAG Family Foundation (ACGIH Foundation for Occupational Health and Safety) (2020) acquired at https://www.acgih.org/foundation/about-fohs accessed on May 1, 2020.

ACKNOWLEDGMENTS

I would like to thank my fellow collaborators on this journey, without whom this initiative would not have materialized. I am forever grateful to Garrett Brown, Perry Gottesfeld Karen Gunderson, Nina Townsend, David Zalk, Ted Zellers, Muhammad Akram, Damien Andrew, Barbara Cohrssen, Noel Correa, Andrew Cutz, Jeff Dalhoff, Chris Laszcz-Davis, Vraj Derodra, Doug Dowis, Dagmar Fung, Jennifer Galvin, Omana George, Dan Grinnell, Stephen Hemperly, Mike Horowitz, David Hornung, Rachel Jones, David Kahane, Carina Kouyoumji, Nimmi Kovvali, Neal Lester, Marianne Levitsky, Aristides Medard, Maharshi Mehta, Sanjiv Pandita, Justine Parker, Jagdish Patel, Diana Peroni, Lydia Renton, Jas Singh, Natalia Varshavsky, Steven Verpaele, Novi Wong, MHSSN, OK International, WHWB/WHWB-US, AIHA National, AIHA Microgrants Subcommittee, Belgian Center for Occupational Hygiene, Forensic Analytical, SGS Galson, SKC Inc., International Safety Systems, and all others that have generously donated their time and resources to this endeavor.

Appendix 5.1
2018/2019 Grants for Trainings and Technical Projects

Developing World Outreach Initiative (DWOI) and Maquiladora Health & Safety Support Network (MHSSN)

Training Grants

Organization/Country	Target Population/Topic
CUIDAR-WIEGO/Brazil	Waste pickers – general OHS
Cividep/India	Tea plantation workers – general OHS
Environics/India	Stonecutters – legal rights
JRP/India	Social Health Activists – general OHS
ICRT – LIPS/Indonesia	Electronics workers – chemical hazards
LION/Indonesia	Chemical workers – chemical hazards
CETIEN/Mexico	Electronics workers – chemical hazards
CFO/Mexico	Maquila workers – chemical hazards
CEISA/Mozambique	University – hazard evaluation/controls
IOHSAD/Philippines	Manufacturing workers – legal rights
WAC/Philippines	Manufacturing workers – general OHS
American University Beirut, Lebanon	Exposure Assessment Basics Training
Karnali Academy of Health Sciences, Jumla, Nepal	Training on Occupational Health and Safety (OHS) to Undergraduate Students of Public Health and Nursing

TECHNICAL PROJECTS

Organization/Country	Target Population/Topic
ICRT – LIPS/Indonesia	Electronics workers – biomonitoring
CEJAD/Kenya	Artisanal gold miners – mercury hazard reduction
Environics, India	Environment Monitoring for an Asbestos Plant, Capacity Building and Training

Appendix 5.2
2019/2020 Grants for Trainings and Technical Projects

Developing World Outreach Initiative (DWOI), Maquiladora Health & Safety Support Network (MHSSN), and Workplace Health Without Borders (WHWB)

TRAINING GRANTS

Organization/Country	Funding Source	Target Population/Topic
MSF-WHWB/Bangladesh	WHWB	Scholarships for OHTA course/BOHS examination
Cividep/India	MHSSN	Women shoe workers – hazard research
Mallarpur Uthnau/India	WHWB	Silicosis – community training
Mallarpur Uthnau/India	DWOI	Silicosis – worker training
CEREAL/Mexico	DWOI	Women electronics workers – OHS/chemical hazards
CETIEN/Mexico	DWOI	Electronics workers – OHS/chemical hazards
CFO/Mexico	MHSSN	Women Maquila workers – gender violence and laws
CEISA/Mozambique	WHWB	University OSH program training
IOHSAD/Philippines	MHSSN	Workers' mental health issues
FTZ-GSE Union/Sri Lanka	MHSSN	Free Trade Zone workers – ergonomics
KKU/Thailand	DWOI	Support for university short course on OHS
CDI/Vietnam	DWOI	Electronics workers – OHS/chemical hazards

TECHNICAL PROJECTS

Organization/Country	Funding Source	Target Population/Topic
Cividep/India	DWOI	Tea plantation hazard research – cash grant
LION/Indonesia	DWOI/Forensic Analytical	Asbestos bulk sampling – cash and donation
JRP/India	DWOI	Silicosis/TB – cash grant
American University/Lebanon	DWOI/SGS	Hazard sampling at university – donation
CEPHED/Nepal	DWOI	Mercury contamination sampling – cash grant

Appendix 5.3
Distribution List of NGOs and Universities for DWOI Requests for Project Proposals

American Friends Service Committee (AFSC)

ANROEV

As You Sow

Asia Coordinator, World Solidarity

Asia Floor Wage Alliance

Asia Monitor Resource Center

Asia Pacific Workers Solidarity Links

Asian Labour Exchange

Asian Pacific Workers Solidarity

Association for the Rights of Industrial Accident Victims (ARIAV)

Association of Solid and Industrializable Waste Workers of the State of Guanajuato "Lázaro Cárdenas del Río" AC

Australia Asia Worker Links

Building and Woodworkers International (BWI)

Business-Human Rights.org

Cambodian Labour Confederation (CLC)

Center for Environmental Justice and Development

Center for Labor Reflection and Action and member of the Steering Committee of the international network GoodElectronics

Center for Trade Union and Human Rights

Centre de Recherche et d'Education pour le Développement (CREPD)

Centre for Research on Multinational Corporations (SOMO)

Centro de Información para Trabajadoras y Trabajadores A.C. (CITTAC)

Centro de Investigación Laboral y Asesoría Sindical (CILAS)

Centro de Reflexión y Acción Laboral (CEREAL)

CFO Maquiladoras

Chinese Working Women Network

Chongqing Zhongxian Ziqiang Disabled Service Station?

CITTAC, Centro de Información para Trabajadoras y Trabajadores A.C. (Workers' Information Center)

Cividep India – Workers' Rights and Corporate Accountability

Clean Clothes Campaign

CLSN

Coalition of Cambodian Apparel Workers' Democratic Union

Columbia Lighthouse for the Blind

Doughty Street Chambers

(Continued)

Education for the Worker-Driven Social Responsibility Network (WSRN) (Previous Director of
  Communications for the Worker Rights Consortium)
Electronics Watch
Environics Trust
Environmental Health Coalition (EHC)
ETI – Ethical Trading Initiative
Fellow Guilders
Globalization Monitor
GoodElectronics Network
Hazards Magazine
Healthy Supply Chains Initiative
Homeworking Worldwide mapping project in Mexico
Hong Kong University of Science and Technology
Hong Kong Women Workers' Association
Hong Kong Workers' Health Centre
Ilima-Kenya
ILRF – International Labor Rights Forum
Industrial Workers of the World
IndustriALL Global Union
IndustriALL Union
International Campaign for Responsible Technology
International Domestic Workers Federation
International Labor Rights Forum (ILRF)
IOHSAD – Institute for Occupational Health and Safety Development
IPEN
ISSEHS
ITESO – Instituto Tecnológico y de Estudios Superiores de Occidente
ITF – International Transport Workers Federation
ITUC-CSI – International Trade Union Confederation
Jeevan Rekha Parishad – Lifeline Council
KASBI trade union federation
Korean Labor Researcher and Author
Labor Action China (LAC)
Labor Education and Service Network (LESN) (NGO)
LaborNotes
LabourFile publication
Legal Action Center
LION
Mallarpur Uthnau (NGO)
Maquila Solidarity Network
Maquila Tijuana San Diego
Mine Labour Protection Trust
Ollin Calli Cooperative
Open University of Tanzania
Peoples Training and Resource Center
Purdue University – School of Health Sciences
Red de Seguridad y Salud en el Trabajo

(Continued)

Rosalux – Rosa Luxemburg Foundation
Seed
SHARPS – Supporters for the Health And Rights of People in the Semiconductor industry
Sheila Pantry Associates Ltd
Shih Hsin University
SIGTUR – Southern Initiative on Globalisation and Trade Union Rights
SJ Social
Society for DISHA
Solidarity Center
Solidarity for Worker Health (NGO)
SOMO – Centre for Research on Multinational Corporations
The Hazards Campaign
Time Up Thailand publication
TLIEA – Taiwan Labor Information & Education Association
UABCS – Universidad Autónoma de Baja California Sur
Universidad Autónoma Metropolitana, Unidad Xochimilco
Universiti Sains Malaysia
University of Sterling (Scotland)
University of the Philippines
USW – United Steelworkers Union
WAC – Workers' Assistance Center Inc.
WHO – World Health Organization
Worker Empowerment (DWOI) labor organization
Worker Rights Consortium
WorkersRights.Org
Worksafe
Zi Ting – Sexworker Advocacy

6 International Education, Research, and Service Opportunities for Students and Faculty in Higher Learning Institutions

Steven M. Thygerson
Brigham Young University

CONTENTS

6.1 GLOBAL OCCUPATIONAL HEALTH

According to the World Health Organization (WHO), only one-third of all countries have governmental programs addressing occupational health and only 15% of all workers globally have access to preventive occupational health services. These services include health surveillance, training in safe work practices, first aid, and employer safety and health programs. WHO's work to provide access to proven occupational health interventions and services includes "stimulating international efforts to build human resource capacities for workers' health, both in primary care and occupational health specialists" (WHO, 2007).

89

According to the National Institute for Occupational Safety and Health (NIOSH) located in the USA, building occupational health capacity in international settings consists of the following:

1. Sharing information and research globally,
2. Contributing to international documents,
3. Providing international technical assistance,
4. Participating in international committees and professional associations, and
5. Contributing to international training materials (NIOSH, 2017).

Building capacity in occupational hygiene will always be a challenge given that this field is much less known when compared to other healthcare and public health fields. In the USA and Europe, students entering universities rarely know about this particular field of public health and declaring to be a student studying occupational health is rare in the first few semesters. This field of study is usually discovered after some chance encounter leading them to occupational health practice. As building capacity in occupational health is even waning in North America and Europe, it is so much more the case in developing countries. For example, Nepal, a country known for its mountains, tourism, and humanitarian opportunities, currently has one occupational health professional trained at the doctoral level, one occupational physician, and less than ten individuals with an earned occupational health master's degree. This may be a similar story in other developing nations.

How do we attract students to be the next occupational hygienists? This chapter will describe how students and faculty from across the globe can be involved in research, education, and service in the exciting field of occupational hygiene. Recommendations will be given regarding how to get involved in these activities. While this chapter focuses on several academic activities, this also applies to the ideas, interests, and skills from occupational health professionals outside of academia who can transfer their knowledge and skills to needed areas. Challenges and rewards by being involved will be described through specific projects and programs that have been undertaken to build occupational health capacity worldwide.

6.2   FACULTY AND STUDENT RESEARCH
IN DEVELOPING COUNTRIES

Research, education, and training are needed for OHS professionals, workers, and employers to address global OHS issues and their local impact (Lucchini & London, 2014). University and college faculty responsible for maintaining occupational health research agendas typically focus on domestic issues ranging from silica exposure among miners to ergonomic-related injuries for construction workers. When researchers develop a passion and desire to expand their research agendas to the developing world, focusing on one's current research interests is a key starting point.

Global health is "the area of study, research, and practice that places a priority on improving health and achieving equity in health for all people worldwide" (Koplan, Bond & Merson, 2009). Global occupational health can be defined as applying occupational health strategies at the global scale, in other words, using the science and art of occupational health to gather and expand knowledge to the various industrial and cottage industry settings around the world. To understand how occupational health professionals can be involved in education, research, and service opportunities in developing countries, a brief history of several projects will be presented.

6.2.1 NEPAL BRICK KILNS

Nepal's brickmaking industry lacks occupational health services. Child labor, bonded labor, and hazardous working conditions are still prevalent in this industry. Improving brick kiln working conditions is now garnering the interest of organizations such as the American Industrial Hygiene Association (AIHA), Workplace Health Without Borders (WHWB), and Global Fairness Initiative, Inc. (GFI). The local academic institutions Kathmandu University and Kathmandu College of Medical Sciences have collaborated with AIHA, WHWB, GFI, and others with the objective of recognizing the hazards and poor working conditions that exist in the brick kilns. Once hazards are recognized, feasible and acceptable controls can be implemented.

Since 2008, Dr. William Carter, professor emeritus at the University of Findlay, has worked with Kathmandu University on various occupational health research projects. Dr. Carter received a core Fulbright research award in 2009. He worked with two master's students in the study of solid waste management. As part of this work, Dr. Carter visited several brick kilns in Bhaktapur (5–6 km east of Kathmandu) where he learned more about the brickmaking process and the hazards in the kilns. Dr. Carter and his research collaborators published two studies during this time on dust and noise exposures for municipal workers and traffic police in Banepa, Nepal (Carter & Rauniyar, 2011; Majumder, Islam, Bajracharya & Carter, 2012). He also published an important work entitled "Introducing Occupational Health in an Emerging Economy: A Nepal Experience" (Carter, 2010). He worked closely with Kathmandu University and Dr. Sanjay Nath Khanal, a professor in the Environmental Science and Engineering Department of Kathmandu University. Dr. Khanal introduced Dr. Carter to a new doctoral student named Seshananda Sanjel.

Seshananda Sanjel received a master's degree in epidemiology and desired to continue his education at the doctoral level. His research mentor suggested he work closely with researchers from the USA by conducting silica and particulate matter sampling research in the brick kilns. Seshananda had limited training in developing sampling protocol for silica and particulates. However, he possessed a wealth of experience in statistics, epidemiology, and research methodology. His willingness to work with research mentors, along with his desire to improve working conditions in Nepal, proved to be a winning formula.

Dr. Steven M. Thygerson, associate professor at Brigham Young University, met Dr. Carter after the 2014 AIHce conference in San Antonio, Texas. Dr. Carter was a committee member for the AIHA International Affairs Committee (IAC) and the committee chair for the brick kiln committee of WHWB. Marianne Lewitsky introduced these two researchers. Immediately, they began to collaborate on how to improve occupational health capacity in Nepal. Both became mentors for Seshananda as he developed his dissertation protocol for silica and particulate sampling. In January of 2015, Dr. Thygerson arrived in Nepal to work extensively with Seshananda and his research team consisting of two master's-level students in environmental health. Dr. Thygerson was tasked to train each team member in appropriate sampling methods, instrumentation, calibration, data acquisition, subsequent laboratory analysis, statistical analysis, and report writing. Through academic research funding, Dr. Thygerson was able to donate several sampling pumps, provide use of a real-time dust monitor, and fund the payment of the silica and particulate laboratory analysis. The results of the study showed that workers were exposed to silica concentrations 17 times greater than occupational exposure limits (Sanjel et al., 2018). Seshananda, Dr. Carter, and Dr. Thygerson continued their collaboration resulting in six peer-reviewed journal articles, five international presentations which culminated in receiving the 2018 International Research Collaboration Award from the International Occupational Hygiene Association (IOHA).

In 2016, Kathmandu University selected Dr. Thygerson to be a Fulbright Specialist focusing on education of environmental health students, nursing students, and medical students. These students received 5 weeks of occupational health training as part of their already rigorous study program. Kathmandu University desired to have more undergraduate- and graduate-level students receive training in occupational health practices including the evaluation of hazards through air and noise sampling. In 2018, Dr. Thygerson and Dr. James Johnston, also from Brigham Young University (BYU), traveled to Nepal to continue brick kiln research. A major objective of this work was to involve the Kathmandu University students in the research process. BYU students were trained on industrial hygiene sampling techniques prior to the trip in order to mentor local KU students in air sampling protocols (see Figure 6.1).

This proved to be a successful collaboration and has continued into its third year. Students are now involved in brick kiln improvement research. Future work will involve engineering students from both Nepal and the USA working to control exposures to silica during the red and green brick carrying tasks. These capstone experiences will involve several more faculty and many more students.

6.2.2   OCCUPATIONAL HEALTH TRAINING IN MOZAMBIQUE

Mozambique is a beautiful country located along the southeastern coast of the African continent. People are attracted to Mozambique by its stunning beaches and local seafood cuisine. Its economy is bolstered mostly by agriculture. However, industry is growing and the need for occupational health and safety priorities is ever needed (CEISA, 2005). Maputo, the capital city of Mozambique, is home to the University of Eduardo Mondlane (UEM). Center for Industrial Studies, Safety and Environment (CEISA) is housed in this university and acts as the research and risk management group for occupational health in Mozambique. CEISA conducts several research projects focusing on the informal industry sector. This includes stone crushing, metal fabrication, sawmills, and brickmaking. In addition, CEISA works directly with professors at UEM to ensure courses integrate workplace safety and health in the curriculum.

In order to provide a broad understanding of occupational health and safety, CEISA and UEM invited the non-profit organization Workplace Health Without

FIGURE 6.1 BYU student teaching silica sampling techniques to Kathmandu University students.

Borders (WHWB) to send two instructors for a week-long course taught on UEM's campus. Custodio Muianga who works for the Agency for Toxic Substances and Disease Registry (ATSDR) in the USA is a native of Mozambique and has worked with CEISA on several projects. He, along with Dr. Steven Thygerson, provided the week-long certificate course in occupational safety and health. Funding for the instructors' travel costs was provided by Underwriters Laboratory. The instructors received no compensation for time or travel. Participants included faculty and staff representatives from each academic department on campus. CEISA employees also attended the course.

Prior to the course, both instructors toured and inspected several classrooms, academic departments, and maintenance areas, which informed some of the risk assessment activities covered during the course. Air sampling methods were a major part of the course. Participants received hands-on instruction regarding integrated sampling equipment including direct reading instruments. Participants used solid sorbent tubes, noise dosimeters, and calibration equipment for air sampling pumps as part of the instruction. The final day consisted of participants conducting hazard recognition, evaluation, and control in their own academic departments.

The purpose of the course was to educate faculty and staff about the occupational health and safety hazards that exist in the industries for which they are training future graduates to enter. At the same time, faculty and staff are instructed about the hazards and control implementation for their respective research areas. Thirty-three participants received certificates for participating in the course.

6.3 OCCUPATIONAL HEALTH CURRICULUM DEVELOPMENT AT INTERNATIONAL UNIVERSITIES

The United Nations General Assembly (UNGA), the International Labor Organization (ILO), WHO, the International Commission on Occupational Health (ICOH), and the European Union (EU) encourage all countries to develop occupational health services for all working people throughout the world. These organizations have identified universities as playing a major role in providing the necessary occupational health services. Yet, the ability for universities to provide these necessary services may be severely lacking (Rantanen, et al., 2017).

Some university programs have entire degrees dedicated to occupational health; others, like engineering or construction management, may have one to two courses to provide students with a background in OHS. Countries with stronger economies, such as South Africa, may have several universities providing degreed programs in OHS. However, there is a need for adherence to an international education standard such as that provided by accrediting organizations such as ABET or Environmental Health Accreditation Council (EHAC). These accrediting bodies are based in the USA, and hundreds of university programs (engineering, computer science, occupational health, and safety) are accredited by them. ABET, for example, will conduct accreditation reviews for programs outside the USA but only under the express and explicit permission from all of that country's national education authorities (ABET, 2020). This can be a very time-consuming, onerous task to obtain permissions, and in many cases, it will not be granted. Programs seeking to improve OHS curriculum without accreditation can still use the accreditation criteria to map out and manage their programs. However, the program directors at these universities may need help assessing and interpreting the information. This is where OHS professionals and university OHS program directors from the US could provide assistance to faculty in other countries. The following section provides an example of such guidance to a new OHS program outside the USA.

6.3.1 THE UNIVERSITY OF LIMPOPO, SOUTH AFRICA

South Africa has eight universities offering degrees or courses in OHS. The University of Limpopo started an OHS program in 2005 and is currently under the direction of Professor Karlien Linde. Professor Linde obtained her training from North-West University and was hired to improve the OSH program at the University of Limpopo. The university supported her in these efforts by allowing consultants from several organizations to take part in a 3-day curriculum development workshop. Representatives from the University of Limpopo (professors, lecturers, advisors), from other South African universities, and from the OHS professional organization

Southern African Institute for Occupational Hygiene (SAIOH) attended with the responsibility to align OSH curriculum with national accrediting bodies.

Dr. Steven Thygerson was invited as part of the US Fulbright Specialist program. His role was to bring international experience in terms of necessary OSH curriculum. Dr. Thygerson spent 3 weeks working at the University of Limpopo with principal responsibilities for curriculum development and adherence to international accrediting bodies such as ABET and EHAC. While in South Africa, he also taught several courses to environmental/occupational health master's students.

Evidence of success after the 3-day meeting was a curriculum map of a 4-year OHS degree that meets university and South African Qualifications Authority (SAQA) accreditation criteria. SAQA recognizes SAIOH as the Professional Registration Body for Occupational Hygiene in South Africa. SAIOH then registers and recognizes candidates for certification as either an Occupational Hygiene Assistant (OHA), Occupational Hygiene Technologist (OHT), or Occupational Hygienist (OH). To receive recognition of the professional status in South Africa, SAIOH will assess, certify, and register all candidates (SAIOH, 2019). Therefore, it is imperative that any program teaching occupational health in South Africa meet the accreditation criteria. Table 6.1 shows the program curriculum that was developed during the workshop. The University of Limpopo expects to have the curriculum fully implemented in the next 3 years with accreditation following shortly after.

This collaboration of OHS professionals with academia demonstrates another way to be involved in building occupational health capacity throughout the world. Fulbright projects for OHS professionals in the USA have been mentioned several times in this chapter as a means to provide research, capacity building, and service in developing countries. To learn more about the Fulbright opportunities, visit https://fulbrightspecialist.worldlearning.org/ for the Fulbright Specialist Program and https://fulbrightspecialist.worldlearning.org/ for the Fulbright Scholar Program.

TABLE 6.1
Proposed Curriculum Map for Occupational Hygiene Degree at University of Limpopo

Name of Module	Short Description of Content
	First Year
Introduction to life sciences I	Homeostasis, cell structure, etc.
Math	Algebra
Introduction to occupational and environmental health	Basic principles of occupational and environmental health
Chemistry	General, organic, inorganic, physical, and analytical chemistry
Physics	Physics relevant to occupational and environmental health
Statistics	Basic principles of statistics
Scientific communication	Basic language skills
Introduction to life sciences II	Systems, etc.

(Continued)

TABLE 6.1 (*Continued*)
Proposed Curriculum Map for Occupational Hygiene Degree at University of Limpopo

Name of Module	Short Description of Content
	Second Year
Introduction to occupational legislation and ethics	South African and international occupational hygiene legislation and ethics
Ergonomics	Ergonomics principles, anthropometry, and human factors
Sociology	Basic sociological concepts and community development
Systems physiology	Physiology of all the major systems in the human body
Scientific writing	Journal articles, PowerPoint presentations, and Excel
Epidemiology	Basic principles of epidemiology
Extreme temperature and pressure	Physics, exposure, evaluation, and control of extreme temperatures and pressure
Radiation: ionizing and non-ionizing	Physics, exposure, evaluation, and control of both ionizing and non-ionizing radiation
	Third Year
Noise	Physics, measurement, and control of exposure to noise
Toxicology A	Toxicology theory and application in occupational setting
Environmental pollution	Air, water, and soil pollution as well as waste handling
Biostatistics and research methodology	Introduction to biostatistics and research methodology
Scientific writing	Report writing, Turnitin/plagiarism, referencing
Hazardous chemical substances A	Measurements of hazardous chemical substances exposure
Hazardous biological agents	Measurement and control of exposure to hazardous biological agents
	Fourth Year
Vibration	The physics of vibration, measurement and control of exposure to vibration
Hazardous chemical substances B	The measurement and control of exposure to hazardous chemical substances
Scientific communication	Writing of reports
Research project	Biostatistics and research methodology lectures and mini-dissertation
Risk assessment	Theory and practice of occupational hygiene risk assessment
Aerosols/fibrogenic dust	Physics and chemistry of exposure to aerosols/fibrogenic dust as well as measurement

6.3.2 Kathmandu University, Nepal

Kathmandu University (KU) in Dhulikhel, Nepal, sits approximately 20 km east of Kathmandu, the capital of Nepal. KU was established in 1991 with strong science, engineering, and technology programs. The environmental science degree is one such program started by Dr. Sanjay Nath Khanal. Under his direction, KU became a regional leader in producing graduates trained in the use and conservation of natural resources while providing some assurances of environmental protection.

With age-old industries such as the brickmaking factories and carpet industry, occupational health was beginning to be seen as a necessity that industries or the government were not providing. As such, Dr. Khanal invited Dr. William Carter in 2009 to assist in developing occupational health curriculum at KU. During his time as a US Fulbright Scholar, Dr. Carter not only conducted research, but also worked to develop the initial curriculum for the university OHS program. Graduate and undergraduate students could now enroll in courses covering the recognition, evaluation, and control of workplace hazards. Students were heavily involved in the silica exposure research conducted in the brick kilns. Students were responsible for conducting questionnaires and the actual silica sampling.

After Dr. Carter left Nepal, the program remained strong. KU continued to invite visiting OHS professors to teach courses or one-time lectures to students while maintaining the strong research studies occurring in the brick kilns. Dr. Seshananda Sanjel is a graduate of the KU OHS program. He is now the current director of the environmental/occupational health program at Karnali Academy for Health Sciences in Jumla, Nepal, a program he started in 2018.

6.4 CHALLENGES TO INTERNATIONAL EDUCATION, RESEARCH, AND SERVICE OPPORTUNITIES

There are obvious challenges when working with international institutions of higher learning. The challenges discussed in this section focus on OHS program issues (equipment, scheduling, students) and not necessarily on cultural competencies. Culture competencies are beyond the scope of this chapter. However, when working with international universities and personnel, local customs should be identified and followed in order to have the best experience possible.

When teaching, conducting research, and providing service in the OHS profession, the use of equipment (sampling pumps, calibrators, noise instrumentation, etc.) is often necessary. Challenges in shipping equipment, bringing equipment with you on a plane, or relying on the local university equipment are numerous. Teaching students of a different culture has several unique challenges. Working in field locations introduces an entirely new set of challenges when working abroad. Table 6.2 below is a short list of challenges and possible solutions to deal with these issues in international settings. Each international setting is unique, and the challenges listed do not represent an exhaustive list of all the challenges one could face.

6.5 CONCLUSIONS

International travel is one of the many reasons to work with international organizations and especially universities. The more you do it, the better you become at being prepared, having better experiences, and meeting new people who become lifelong friends. No matter what stage you are at in your OHS career, look into international education, research, and service opportunities. Your small, simple contributions make very big differences.

TABLE 6.2
Challenges and Solutions When Teaching, Researching, or Providing Services in International Locations

Challenge	Possible Solution
	Equipment

Clearing customs and airport security – items such as pumps, tubing, wires, sample media may look suspicious to transportation security personnel.	1. Follow all airport security regulations pertaining to the country which you are visiting. Visit www.tsa.gov for information regarding traveling to international airports via the USA.
	2. Obtain an official letter from both your institution and the international institution stating the following:
	  a. Who is bringing the equipment into the country and who in the country is responsible for the equipment.
	  b. What is the equipment (make, model, serial number for each piece).
	  c. Where the equipment will be used.
	  d. When the equipment will be used and when the equipment will return.
	  e. Why the equipment is being used. What is the specific purpose?
	Keep a copy of this letter with you and in the baggage of the equipment where it is easily identified by security personnel.

Transporting equipment to prevent theft and damage	1. Be judicious in selecting equipment. Is each instrument truly necessary? 2. Use a reputable international courier company (FedEx, UPS) to ship equipment. Insurance should be purchased, and signatures of receipt should be required. However, this can be very expensive to ensure the equipment arrives safely and on time. 3. Use carry-on baggage if possible for the most expensive items. 4. All lithium batteries must be in carry-on baggage. 5. Check the equipment as part of your checked baggage. This cost-effective option should be considered first as most international long-haul flights allow extra baggage per traveler. Ensure all equipment is well protected to prevent damage.
Equipment electrical power requirements	1. Determine the electrical power requirements prior to the trip. DO NOT plug in equipment to charge or operate unless you know the power requirements for each instrument. 2. Purchase electrical converters and adapters in your home country, if necessary rather than waiting until in-country. 3. Use equipment requiring AA or AAA batteries as these batteries are readily accessible in most countries.
In-country maintenance of equipment	1. Perform all preventive maintenance only as suggested by equipment manufacturers. 2. Bring spare parts (filters, tubing) recommended by the manufacturer. 3. Avoid in-country maintenance and repair shops. They are unfamiliar with the equipment, may cause further damage, and their work may void any manufacturer warranties. 4. If you are leaving equipment in-country, teach in-country colleagues how to conduct the preventive maintenance recommended by the manufacturer. DO NOT allow in-country colleagues to perform any other equipment maintenance. Have them send the equipment back to you or leave it for your next trip.

(Continued)

TABLE 6.2 (*Continued*)

Challenges and Solutions When Teaching, Researching, or Providing Services in International Locations

Challenge	Possible Solution
	Teaching
Cultural differences in teaching, learning, assessment, and classroom protocol.	1. Discuss with your host-specific classroom protocol (attendance, note-taking, excuses, entering/exiting the classroom, discussions versus lecture).
	2. Understand how students learn in the country you are visiting (visual, kinesthetic, memorization, etc.). For example, many students are accustomed to simply listening to lectures and memorizing information. They may not be used to or prepared for discussions or any type of talking in class. Be prepared for this but also use innovative teaching techniques – this is why you were invited to teach in the first place.
	3. Learn the types of assessment activities the university uses (quizzes, examinations, open or closed book, oral examinations). Be prepared for this but also use innovative methods to assess student learning.
	Service and Field Work
Constant changes to your schedule	1. Be prepared for Murphy's law to be in full effect. If something can happen, it most likely will.
	2. Be patient with all schedule changes. You are working on their time, not yours.
	3. Do not overschedule each day. If necessary, call the day short and enjoy the country.
	4. Make daily plans knowing things willchange.

REFERENCES

ABET. (2020). Accreditation outside the U.S.: ABET's international role. Retrieved March 2020 from https://www.abet.org/global-presence/accreditation-outside-the-u-s/.

Carter, W.S. (2010). Introducing occupational health in an emerging economy: A Nepal experience. *The Annals of Occupational Hygiene*, 54(5), 477–485.

Carter, W.S., Rauniyar, R. (2011). When the exchange rate makes a difference: Noise Monitoring of Traffic Police in the Kathmandu Valley, Nepal. *International Journal of Occupational Health and Safety*, 1(1), 7–13. doi: https://doi.org/10.3126/ijosh.v1i1.4490.

CEISA. (2005). Work and Health in Southern Africa. Situational analysis on informal small scale sector enterprise in Maputo. Durban: Centre for Industrial Studies, Safety and Environment (CEISA) and University of KwaZulu-Natal (UKZN).

Koplan, J.P., Bond, T.C., Merson, M.H, Reddy, K., Rodriguez, M., Sewankambo, N., and Wasserheit, J. (2009). Towards a common definitions of global health. *Lancet*, 373, 1993–1995. doi:10.1016/S0140–6736(09)60332-9.

Lucchini, G.R. and London, L. (2014). Global Occupational Health: Current Challenges and the Need for Urgent Action. *Annals of Global Health*, 80(4), 251–256. doi:10.1016/j.aogh.2014.09.006.

Majumder, A.K., Islam, K.M., Bajracharya, R.M., Carter, W.S. (2012). Assessment of occupational and ambient air quality of traffic police personnel of the Kathmandu valley, Nepal; in view of atmospheric particulate matter concentrations (PM10). *Atmospheric Pollution Research*, 3(1), 132–142.

NIOSH. (2017). Global outreach. The National Institute for Occupational Safety and Health. U.S. Department of Health and Human Services, Washington, DC. Retrieved March 2020 from https://www.cdc.gov/niosh/topics/global/default.html.

Rantanen, J., Lehtinen, S., Valenti, A, Lavicoli, S. (2017). A global survey on occupational health services in selected International Commission on Occupational Health (ICOH) member countries. *BMC Public Health*, 17, 787. doi: 10.1186/s12889-017-4800–z.

SAIOH. (2019). South Africa Institute for Occupational Hygiene. Retrieved December 2019 from https://www.saioh.co.za/.

Sanjel, S., Khanal, S.N., Thygerson, S.M., Carter, W.S., Johnston, J.D., Joshi, S.K. (2018) Exposure to respirable silica among clay brick workers in Kathmandu valley, Nepal. *Archives of Environmental and Occupational Health*, 73(6), 347–350. doi:10.1080/19338244.2017.1420031.

World Health Organization (WHO). (2007) Workers' health: global plan of action. Sixtieth World Health Assembly. Geneva, Switzerland [cited 2020 March 26]. Retrieved from https://www.who.int/occupational_health/WHO_health_assembly_en_web.pdf.

7 The Role of Equipment and Services Vendors in Advancing Occupational Hygiene Globally

Deborah F. Dietrich

CONTENTS

7.1   INTRODUCTION

The pages of this book are filled with examples of how the health and safety of workers globally have been improved through the volunteer efforts of occupational hygiene professionals sharing their time and expertise. Government agencies such as the US National Institute of Occupational Safety and Health (NIOSH) have also contributed to this effort through site visits, training, and collaboration with professionals in targeted regions. Similarly, a tripartite agency of the United Nations, the International Labor Organization (ILO), has advanced occupational hygiene by developing policies, programs, and standards that promote worker health. As with any profession, however, a driver of long-term sustainable growth of the global occupational hygiene profession is the engine of economics. Specifically, vendors of occupational hygiene equipment and services must be willing to make an investment in the region for the profession to thrive.

To illustrate this point, let's first examine one of the fundamental components of an occupational hygiene program – air sampling. To establish a credible air sampling program, occupational hygienists must have access to and training on the selection and use of air sampling equipment. They must also have training on published sampling and analytical methods, applicable regulations, and established occupational exposure limits and guidelines.

Air sampling equipment vendors can help fill that need through their own efforts and through collaboration with others. Air sampling equipment suppliers such as SKC Inc. with a global network of distributors not only sell the essential sampling tools, but also provide users many free training resources. Online webinars, videos, presentations, and sampling guides enable global occupational hygiene practitioners with various levels of expertise to learn about the science and technology of sampling from any location with Internet access.

Scientists employed at corporate locations work in concert with their global distributors to keep them well-informed on technical details of sampling and adequately trained on sampling products. In this way, local vendor representatives speaking the native language have the expertise to not only help users select the required sampling equipment but can show them how to use it and maintain it. Free online technical support on products is also available upon demand from certified industrial hygienists and other scientists. For more personalized support, scientific professionals working for vendors frequently travel to present on-site courses or seminars to share the latest developments with occupational hygiene groups throughout the world. These events not only advance occupational hygiene knowledge, but also create networking opportunities for future information exchange. On an informal level, vendor representatives who travel globally often serve as long-term mentors and friends to those they meet in their travels enhancing professional growth and job satisfaction.

Working through non-profit organizations, equipment vendors also advance occupational hygiene by providing equipment grants and equipment loans for training or other critical needs. The Occupational Hygiene Training Association (OHTA) described in another chapter of this book has developed training materials that are freely available and used by students and training providers throughout the world. For many of these courses, however, hands-on learning with occupational hygiene equipment is necessary to ensure a complete grasp of the science. For example, air sampling equipment is an essential part of the learning experience for the OHTA W501 course on Measurement of Hazardous Substances described at OHlearning. com. But buying and shipping equipment all over the world for training courses would be overly burdensome on the budget of OHTA and volunteers.

Equipment vendors step in to fill this vital role. SKC, for example, maintains W501 sampling kits at their offices in the USA, UK, and Singapore packed with all the necessary supplies for the course. Upon request from the designated training provider, the items are shipped without cost and vendor representatives often attend the course to help train students to use the equipment.

Similarly, vendors collaborate with organizations such as Workplace Health Without Borders and the Developing World Outreach Initiative described in another chapter of this book to provide equipment grants. The equipment that is provided by vendors for training and special projects helps to address critical concerns in developing countries and regions of the world with underserved worker populations. Some examples of recent equipment grants include pumps, calibrators, and filters for asbestos sampling in India and passive samplers for anesthetic waste gas and VOC sampling in Lebanon. Vendors also provide equipment grants to international universities offering degree programs in occupational hygiene or related fields. For example, in 2012, a generous donation of air sampling equipment was given to a university in Indonesia helping students learn the art and science of occupational hygiene using the latest technology. Equipment grants also provide students the opportunity to become familiar with key vendors, products, and services for use in their jobs.

Commercial laboratories also play a key role in the growth and sustainability of occupational hygiene globally. Occupational hygienists cannot do their jobs without access to qualified analytical laboratories providing a full range of air sample analysis services. Even if air samples must be sent back to the USA or UK for analysis, some laboratory agencies such as SGS and Analytics Corporation have established a presence in international locations to provide critical support services such as pump loan, technical support on methods and media, along with consulting and auditing services. Similarly, the presence of ISO 17025 calibration laboratories in the region allows occupational hygienists to certify flowmeters, noise calibrators, and other sampling instruments annually to national standards in a convenient and cost-effective manner.

Vendors serving the global occupational hygiene community not only have a desire to protect workers, they have an economic incentive to do so. Economists refer to the "invisible hand" that drives supply and demand through a free market. This invisible hand metaphor was first introduced by Adam Smith, an eighteenth-century Scottish philosopher and economist, to characterize the mechanism through which beneficial social and economic outcomes arise from the accumulated self-interested actions of individuals (Smith, 1759). In the examples above, you can see that there is indeed a beneficial outcome for all parties when vendors of equipment and services commit their time, energy, and expertise to the development of occupational hygiene globally.

Once the commitment and passion of practitioners are combined with the products, services, and training available from vendors, the occupational hygiene profession is positioned for growth in the region. The combined forces of practitioners and vendors typically lead to the development of occupational hygiene associations. Associations provide the platform for an occupational hygiene movement in the region. Through the association, occupational hygienists overcome the feeling of working in isolation. The association provides a forum for networking, mentoring, and continuing education to keep hygienists on the cutting edge of the science of occupational hygiene.

FIGURE 7.1 Vendor representatives from the Asia-Pacific Region received training on the operation, maintenance, and repair of air sampling pumps from corporate representatives to better serve industrial hygienists in their home country. (Photo courtesy of Deborah F. Dietrich.)

Vendors provide necessary funds through sponsorships and tradeshows that enable the association to grow and offer its members courses and other events. Vendor advertisement dollars also spur the creation of association journals and other technical publications. Finally, vendors often provide loaner equipment, technical experts at international conferences, and a variety of training resources. Figure 7.1 shows a vendor teaching a course on air monitoring in the Asia-Pacific Region.

In the end, the global occupational hygiene professional often grows initially through the seeds planted by volunteers and agencies outside of the region. These efforts pave the way for vendors to enter a new market in what will hopefully be a mutually beneficial relationship that results in highly capable occupational hygiene professionals and better protection of worker health. The following case studies illustrate the synergy that results when vendors working with other professionals commit to the growth of occupational hygiene in a targeted region.

7.2 THE VIETNAM EXPERIENCE

7.2.1 WORKFORCE AND WORKPLACES

With a population near 97 million in the year 2020, Vietnam is the 15th most populous country in the world. The median age in Vietnam was 32.5 years, and the number of employed persons was 55 million at the end of 2019 (Worldometer, 2020c; Trading Economics, 2020a). At a technical session presented at the 1998 American Industrial Hygiene Conference and Exposition (AIHCE), NIOSH representatives reported that 90% of all Vietnam workers at that time were engaged in agriculture as part of an informal economy (Waters, 1998). As of 2020, Vietnam is still a leading agriculture exporter and food processing hub. But Vietnam is now the 47th largest economy in the world with industries such as electronics, manufacturing, mining, finance/banking, and information technology. Vietnam is also one of the largest oil producers in Southeast Asia, and 48 Vietnamese companies were named the best workplaces in Asia in 2019 (Kiprop, 2018; Trading Economics, 2020b; Nguyen, 2019).

7.2.2 Occupational Health and Safety Regulations and Governing Authorities

The Vietnam National Assembly passed a milestone occupational safety and health law in June 2015. This new law was drafted following The Promotional Framework for Occupational Safety and Health Convention ratified at the 2006 general conference of the ILO and was applauded for extending protection and preventive efforts to the informal economy in this country (International Labor Organization, 2006; International Labor Organization, 2015).

The Ministry of Labor, Invalids, and Social Affairs (MOLISA) is the institution that manages work-related issues in Vietnam. MOLISA Labor Inspectors are responsible for ensuring that employers comply with the legislation in regard to general working conditions, payment of social security dues, establishment of trade unions and collective bargaining agreements, and handling of labor disputes. National MOLISA inspectors or alternatively Provincial Departments of Labor, Invalids, and Social Affairs (DOLISAs) may also take on an advisory role providing technical information and assistance to employers and employees alike (International Labor Organization, 1996–2020). Like many countries, however, the Labor Inspectorate is much more focused on enforcement efforts than on preventative efforts.

Workplace inspections are based on an annual inspection plan, and employers are given 2–3 days advance notice. Employers are also required to submit a self-assessment report to the Labor Inspectorate. In case of any violations noted by the inspectors, a variety of sanctions are proposed, and the employer has the right to appeal.

7.2.3 International Outreach and Vendor Support

In 2015, the Vietnamese National Institute of Occupational and Environmental Health (NIOEH) signed a Memorandum of Understanding (MOU) with the American Industrial Hygiene Association (AIHA) followed by a Memorandum of Collaboration with the US NIOSH in 2016. A central goal of this collaboration was to fill the knowledge gap in occupational hygiene at such a critical time in history as Vietnam became more industrialized. Other goals included expanding the number of occupational hygiene professionals who were not only trained but had achieved internationally recognized certification. Once trained and certified, these professionals could not only practice their profession but also serve as trainers to help expand the profession in the country. This is very important as language can be a barrier to training efforts when using trainers from outside the region. A final goal of the MOU was to build long-standing relationships with international experts that could be a source of information and collaboration in the future.

As a result, several courses developed by the OHTA have been taught through the years such as the OHTA W201 (Principles of Occupational Hygiene) and OHTA W501 (Measurement of Hazardous Substances). Occupational hygiene experts from

AIHA and other organizations such as Workplace Health Without Borders and the Developing World Outreach Initiative volunteer their time to this effort.

These courses are made more meaningful to students by the generosity of vendors who provide loaner kits at no cost for training purposes. For example, SKC sent loaner kits to Vietnam for W501 courses in Hanoi and Ho Chi Minh, which included air sampling pumps, calibrators, sampling heads and media, and other monitoring equipment. In addition to loaning the equipment, SKC and other vendors provided sponsorship funds and speakers for international conferences in Vietnam such as the Fifth International Conference on Occupational and Environmental Health held in Hanoi in September 2018.

As of the year 2020, the profession of occupational hygiene in Vietnam is well positioned to grow because of the combined efforts of many. The success evidenced thus far is due in part to vendors of equipment and services who have committed resources to this region. Leading manufacturers of occupational hygiene equipment such as Honeywell, Draeger, 3M, and DuPont have established branch offices or subsidiaries in Vietnam to supply and support practitioners who need the essential tools such as gas detectors and personal protective equipment. SKC partners with a local distributor in Vietnam, Victory Instrument JSC, to provide sampling equipment and support in the local language. While there are no AIHA-accredited laboratories that have a laboratory on-site in Vietnam, some laboratory companies have established a local presence for other support services. SGS, for example, established an office in Vietnam in 1997 staffed with scientific professionals that do consulting, auditing, field services, and the EDGE Green Building Certification program. They also have a pump loaner program and supply sample media for use by Vietnam professionals.

7.3 THE INDIA EXPERIENCE

7.3.1 WORKFORCE AND WORKPLACES

With a population of approximately 1.37 billion, India is the second most populous country in the world with a median age of 28.4 years (Worldometer, 2020b). A 2019 report indicated that India had the fastest growing trillion-dollar economy in the world and the fifth largest economy overall (Silver, 2019). It is difficult to accurately determine the size of the entire workforce because many are employed in the unorganized sector. But published reports indicate that nearly half of the workers (43.21% in 2019) are employed in agriculture while the others work in factories and service industries (Plecher, 2020). Agriculture and food processing are still among the most important economic drivers along with industries such as iron, steel, chemicals, pharmaceuticals, cement, and mining. Service industries such as banking, finance, software, and information technology also contribute greatly to the country's economy.

7.3.2 OCCUPATIONAL HEALTH AND SAFETY REGULATIONS AND GOVERNING AUTHORITIES

There are two main acts enacted by the federal government to protect the health, safety, and welfare of workers. They are the Factories Act of 1948 (amended in 1987) and the Mines Act of 1952. The Factories Act is quite comprehensive and includes permissible limits of exposure for many chemicals listed in the Second Schedule. The Factories Act is enforced by Labor Department inspectors at the state level. The states are also empowered to add more detail to the framework in the federal law to meet local requirements. If violations of the law are found by inspectors, the manager of the factory can be found guilty of an offense punishable with imprisonment or fine (International Labor Organization NATLEX database, 1948 amended 1987).

7.3.3 International Outreach and Vendor Support in India

The huge industrial growth in India has resulted in a very large workforce exposed to a variety of occupational health and safety hazards. The situation is further complicated by the lack of occupational hygienists in the country who are properly trained to evaluate and control these hazards. An AIHA *Synergist* article indicated that India needs another 5,000–10,000 occupational hygienists to meet the needs of the country. To meet that need, international occupational hygiene professionals collaborated to begin the first master's degree program in India (AIHA, 2016).

A native of India who was trained and working in the USA, Maharshi Mehta, took the lead. Working with the University of Cincinnati, his alma mater, a MOU was signed in 1996 whereby Cincinnati provided the basic course curriculum for Sardar Patel University in India. Maharshi modified the curriculum to meet the local needs, and the University of Cincinnati faculty visited India each year to help teach the course. The program flourished through the assistance of the global occupational hygiene community who donated books and equipment vendors who donated essential scientific instruments and sample media for hands-on training along with a variety of technical resources. Maharshi advised this author that as of 2019, there have been about 270 graduates from this program and 15 CIHs in the country where there were only 2 CIHs 20 years ago.

The Sardar Patel University program has sparked a whole new energy and excitement for occupational hygiene in India that has led to the formation of the Central Industrial Hygiene Association (ciha.in) in August of 2004. Their stated goal is "to educate and influence society to adopt safety, health, and hygiene policies, practices, and procedures that prevent and control occupational health-related problems". Each year, CIHA provides members continuing education opportunities through an annual Conference and Professional Development Course. This would not be financially feasible without the generosity of vendors who act as sponsors. The list of sponsors for their February 2020 conference entitled *Industrial Hygiene: Keys to Reduce Health Risks* includes DuPont, SKC, and SKC's local distributor, Swan Environmental Pvt. Ltd. A scroll through previous conferences listed on the CIHA website shows consistent support from equipment vendors along with Maharshi Meta's consulting firm International Safety Systems Inc. (ISS).

As in Vietnam, OHTA training courses serve as a mechanism to advance the number and expertise of occupational hygienists in India. Private consultants such as ISS have presented the OHTA Training Module W501 on Measurement of Hazardous Substances with the advantage of speaking the native language. SKC has again provided the no-cost loaner kits for this course, and the local SKC distributor, Swan

Environmental, has assisted with the class to help train students on the specifics of each equipment model. Workplace Health Without Borders and Developing World Outreach Initiative have identified serious occupational health concerns such as silica exposures in India's stone crushing. Equipment grants from vendors have enabled air sampling and other critical work to be done initially by volunteers while simultaneously training locals to carry on the work in the future.

The India experience is an inspiring one when you consider how many individuals and organizations including a university gave of their time and resources to lay a firm foundation for occupational hygiene. Similarly, the donations that vendors of equipment and services have made to the India experience have helped to achieve the growth of the profession evidenced thus far. As the profession thrives, new occupational hygiene businesses are attracted to India that will support practitioners and CIHA activities. Shiva Analyticals, for example, has established a state-of-the-art analytical testing laboratory in Bangalore that includes industrial hygiene testing. Draeger has an office in Mumbai, MSA is in West Bengal, Industrial Scientific is in Delhi, and the SKC distributor, Swan Environmental, is headquartered in Hyderabad. The infrastructure of occupational hygiene has been established, and the engine of economics is in place. The profession is positioned to grow in India, and the master's degree programs along with the other training and volunteer efforts will move it forward in the years ahead.

7.4   THE BRAZIL EXPERIENCE

7.4.1   WORKFORCE AND WORKPLACES

With a population of approximately 212 million in the year 2020, Brazil is the sixth most populous country in the world. The median age in Brazil is 33.5 years, and the number of employed persons was 94.5 million at the end of 2019 with 9% employed in agriculture, 20% in industry, and 70% in the service sector (Worldometer, 2020a; CEIC, 2020; Plecher, 2019). The manufacturing sector in Brazil is the third largest in the Americas, and Brazil is the second largest ethanol fuel producer in the world. Outside of the agriculture industry, the largest industries include oil and gas, automobile production, machinery production such as tractors and road equipment, and textiles (Sawe, 2017).

7.4.2   OCCUPATIONAL HEALTH AND SAFETY REGULATIONS AND GOVERNING AUTHORITIES

Brazil has a unique history in regard to occupational health and safety. Brazil was colonized by the Portuguese in the sixteenth century who developed a lucrative sugarcane industry with help from slaves they brought in from Africa. Later, textiles, coffee, and hats were added to their list of exports. Over the years, business owners noted the loss of productivity and profits that occurred when workers became ill, and by 1900, Brazilian doctors began to realize the relationship between work and health.

In 1943, health and safety issues of work were first included in the Consolidation of Labor Laws, which have been amended extensively since then. Chapter 5 of these laws requires all companies to enforce general standards of occupational health and safety, and an additional set of regulatory standards provide additional detail. Each workplace must also have an Internal Commission for Accident Prevention or CIPA. Enforcement is done by the Labor Inspection Secretariat (SIT) of the Ministry of Labor and Employment (MIT) with inspectors located in each of Brazil's 26 states (Warburton, 2014).

In recent years, the Brazilian federal government proposed a significant new initiative called eSocial for the submission of employer and employee data to the government. The eSocial program creates a single platform for employers to electronically send the government all details related to taxes, social security, and labor issues. Once finalized, eSocial will be very important to occupational hygienists in Brazil because they will be required to send the government all details related to health and safety of workers including accidents, environmental risk factors, and special employee retirements for unhealthy and hazardous environments (Justo, 2017). At last update, this portion of eSocial was scheduled to be implemented in 2021. Central to the advancement of occupational health and safety in Brazil is Fundacentro, a research body similar in function to the US NIOSH and the UK Health and Safety Laboratory. Fundacentro's mission is to develop solutions to protect the health of Brazilian workers and to reduce workplace accidents.

7.4.3 INTERNATIONAL OUTREACH AND VENDOR SUPPORT

Historically, Brazil has been very well connected to the global occupational hygiene community. Brazil native, Berenice Goelzer, led occupational hygiene efforts for the World Health Organization (WHO) headquartered in Geneva, Switzerland, for 25 years, and Brazil's Fundacentro is a WHO Collaborating Center. The US NIOSH has also collaborated with Fundacentro to better train workers on the proper selection and use of respiratory protection given the fact that many workers throughout Latin America use respiratory protection daily (NIOSH, 2012).

This large country with an industrialized workforce and trained health and safety professionals has been a prime market for vendors of occupational hygiene equipment and services. 3M, for example, who manufactures respirators and other safety equipment, has established their own offices in Brazil. As such, 3M vendor representatives speaking the native language serve to enhance the reach for respirator training and support throughout Brazil. To ensure the proper fit, vendors of respirator fit-testing equipment such as TSI and OHD have established local distributors and provide training to users through professional conferences, on-site visits, and remote training resources. Similarly, vendors of high-quality air sampling, noise, and other necessary occupational hygiene tools have established local distributors that will enhance the growth of occupational hygiene in Brazil. Vendors not only provide a product. They provide additional "boots on the ground" for training and support. Internet shopping may provide easy access to products from anywhere in the world. But only credible vendors with a commitment to the region ensure users have the proper tools and training for the occupational hygiene task at hand.

As noted in the introduction of this chapter, the combined forces of practitioners and vendors lead to the development of occupational hygiene associations that propel the growth of the profession. This was the case in Brazil. The Brazilian Association of Occupational Hygienists (ABHO.org.br) was founded in 1994. ABHO provides opportunities for networking and information exchange for their members along with technical committees for scientific discussions and many courses for professional development. ABHO represents the individual and collective interests of the occupational hygienists of Brazil in legislative, regulatory, or other issues and offers an occupational hygiene certification for those that meet the requirements and successfully pass an examination (Analytics, 2018).

Like occupational hygiene associations in many countries, ABHO provides the infrastructure on which the profession in Brazil is built. With a typical attendance of several hundred hygienists at the ABHO national conference each year, this event attracts exhibitors from many parts of the world. The fees generated from exhibit booth space and vendor sponsorships help to make the conference a world-class event with simultaneous translation into Portuguese if needed. Typically, vendors contribute to the conference by providing technical speakers who give presentations or full-day courses on select topics such as Size-Selective Particulate Sampling, Vapor Intrusion, and Nanoparticles. Once all the work has been done to translate the slides from English to Portuguese, training is often repeated later at local conferences throughout Brazil by local equipment vendors such as Faster Comercio in Sao Paulo. Technical presentations in Portuguese are also available for free on-demand from vendor websites, and remote technical support is provided through email.

With the added benefit of fun social events and new friendships, some vendors have become regular attendees at the ABHO national conference. The author of this chapter, for example, attended this event for over 25 years and received an award from ABHO in 2005 for promoting the occupational hygiene profession in the region.

As ABHO grew, so did the occupational hygiene profession and the vendors serving this country. The number of commercial occupational hygiene laboratories in particular has seen an incredible growth in recent years. In 2003, Environ Científica Ltda. in the San Bernardo district of Sao Paulo became the first AIHA-accredited laboratory. Environ has since been acquired by SGS and continues to be a leader in the Brazilian occupational hygiene laboratory market. Solutech Comércio e Serviços de Análises Químicas Ltda. has also achieved AIHA accreditation, and Analytics, another AIHA-accredited laboratory, has a team of chemists on staff in Brazil to support local customers with analysis being done at their headquarters in Ashland, Virginia. The presence of accredited laboratories sends a clear message that the occupational hygiene profession has fully arrived in Brazil and it is poised for growth in the future.

7.5 CONCLUSIONS

One of the hallmarks of the occupational hygiene profession is the spirit of cooperation and camaraderie that exists between all those working to improve the health and safety of workers. In this spirit, occupational hygiene practitioners recognize and welcome the contributions made by vendors in advancing the profession and supporting critical projects. In turn, vendors offer practitioners the convenience of a local business from which they can source critical equipment and services. However, vendor products and services are not the end of the benefits to practitioners. Vendors remain an ongoing source of education, training, and technical support as well as a source of revenue that aid in the formation and success of global occupational hygiene associations.

REFERENCES

American Industrial Hygiene Association Synergist (2016). Pole to Pole: India, An Interview with Maharshi Mehta, https://synergist.aiha.org/201611-pole-to-pole-india, accessed February 12, 2020.

Analytics Corp (2018). ABHO and Certification of Occupational Hygienists, https://analytics corp.com/abho-e-certificacao-de-higienistas-ocupacionais/, accessed on February 11, 1020.

CEIC (2020). Brazil Labor Force Participation Rate, https://www.ceicdata.com/en/indicator/brazil/labour-force-participation-rate, accessed February 3, 2020.

International Labor Organization (2006). C187-Promotional Framework for Occupational Safety and Health Convention, NORMLEX Information System on International Labor Standards, https://www.ilo.org/dyn/normlex/en/f?p=NORMLEXPUB:12100:0::NO::P12100_ILO_CODE:C187, accessed January 27, 2020.

International Labor Organization (2015). Occupational Safety and Health Country Profile: Vietnam, https://www.ilo.org/safework/countries/asia/vietnam/lang--en/index.htm, accessed May 6, 2020.

International Labor Organization (1996–2020). Viet Nam, https://www.ilo.org/labadmin/info/WCMS_150920/lang--en/index.htm, accessed January 27, 2020.

International Labor Organization NATLEX database. India Factories Act Second Schedule. (1948, amended 1987), https://www.ilo.org/dyn/natlex/docs/WEBTEXT/32063/64873/E87IND01.htm, accessed February 17, 2020.

Justo, F. (2017). eSocial is Coming-Is your Brazilian Company Ready?, TMF Group Article, https://www.tmf-group.com/en/news-insights/articles/2017/march/esocial-brazil/, accessed on February 4, 2020.

Kiprop, J. (2018). The Biggest Industries in Vietnam, https://www.worldatlas.com/articles/top-biggest-industries-in-vietnam.html, accessed January 23, 2020.

National Institute for Occupational Safety and Health (2012). NIOSH Reaches Abroad to Protect Workers using Personal Protective Equipment, https://www.cdc.gov/niosh/updates/upd-06-20-12.html, accessed February 4, 2020.

Nguyen, T. (2019). 48 Vietnamese Companies Named Best Workplace in Asia, *The Leader*, https://e.theleader.vn/48-vietnamese-companies-named-best-workplace-in-asia-1562831138678.htm, accessed January 23, 2020.

Plecher, H. (2019). Employment by Economic Sector in Brazil, *Statista*, https://www.statista.com/statistics/271042/employment-by-economic-sector-in-brazil/, accessed February 3, 2020.

Plecher, H. (2020). Distribution of the Workforce Across Economic Sectors in India 2019. *Statista*, https://www.statista.com/statistics/271320/distribution-of-the-workforce-across-economic-sectors-in-india/, accessed January 28, 2020.

Sawe, B.E. (2017). World Atlas, The Biggest Industries in Brazil, https://www.worldatlas.com/articles/which-are-the-biggest-industries-in-brazil.html, accessed on February 17, 2020.

Silver, C. (2019). The Top 20 Economies in the World, *Investopedia*, https://www.investopedia.com/insights/worlds-top-economies/, accessed January 29, 2020.

Smith, A. (1759). The Theory of Moral Sentiments. Scotland, https://www.adamsmith.org/the-theory-of-moral-sentiments, accessed May 6, 2020.

Trading Economics (2020a). Vietnam Employed Persons, https://tradingeconomics.com/
    vietnam/employed-persons, accessed January 23, 2020.
Trading Economics (2020b). Vietnam GDP Growth Rate, https://tradingeconomics.com/
    vietnam/gdp-growth, accessed January 23, 2020.
Warburton, C. (2014). Health and Safety in Brazil: BRIC by BRIC, *British Safety Council*, https://
    www.britsafe.org/publications/safety-management-magazine/safety-management-
    magazine/2014/health-and-safety-in-brazil-bric-by-bric/, accessed on February 4, 2020.
Waters, M., Meinhardt, T., Mullan, R., Nguyen, M. (1998). Occupational Hygiene in Vietnam:
    Challenges and Opportunities, https://www.cdc.gov/niosh/nioshtic-2/20043696.html,
    accessed January 22, 2020.
Worldometer (2020a). Brazil Population 1950–2020, https://www.worldometers.info/world-
    population/brazil-population/, accessed May 6, 2020.
Worldometer (2020b). India Population 1950–2020, https://www.worldometers.info/world-
    population/india-population/, accessed January 28, 2020.
Worldometer (2020c). Vietnam Population 1950–2020, https://www.worldometers.info/
    world-population/vietnam-population/, accessed January 23, 2020.

8 Capacity Crashing
The Ongoing Transfer of Dangerous Jobs and Processes to Underdeveloped Regions

Mary O'Reilly
University at Albany School of Public Health
and Workplace Health Without Borders

CONTENTS

8.1 INTRODUCTION

Capacity crashing is facilitated by a lack of capacity building. It is almost the opposite of capacity building. Capacity building has long been a priority for the International Labor Organization, the World Health Organization, and the US National Institute of Occupational Safety and Health, along with many other groups. Capacity building denotes the process by which individuals and organizations obtain, improve, and retain the skills, knowledge, tools, equipment, and other resources needed to survive and thrive.

Capacity crashing, on the other hand, denotes the practice of moving industrial production or processes to countries or locations that do not have the capacity to protect environmental and worker health in the face of the demands created by those industries. Most businesses, however, do not make decisions solely, or even primarily, based on occupational and environmental health impacts. Business decisions most often rely on an assessment of tax structures, banking capabilities, the promise of expanding markets, and ease of doing business in a particular country, among many other factors. Although worker health and safety are not major considerations when deciding where to locate a business or corporation, it makes it all the more important to build occupational and environmental health capacity to protect vulnerable people, communities, and countries before industrialization begins. Figure 8.1 shows a worker in the garment industry in Vietnam painting clothing by hand.

Although some developing countries may not have enacted adequate laws to protect workers and the environment, many have. Many that have adequate laws may not, however, have a structure robust enough, or sufficient numbers of enforcement people, to ensure the laws are implemented. Factories may rely on older and less intrinsically safe equipment, and there may be insufficient supplies of personal protective equipment available for workers. People struggling to house, clothe, and feed themselves are more likely to settle for any job even if it is unsafe. And, unfortunately, every society has a certain number of people willing to profit at the expense of their fellow human beings and fellow citizens.

Another aspect of capacity crashing is risk transfer. Risk transfer is a risk management and control strategy that involves the contractual shifting of risk from one party to another. Large companies and brand name companies transfer occupational risk to smaller companies and supply chains. Despite the emphasis of corporate governance

FIGURE 8.1 Vietnamese shirt painter. (Photo courtesy of Andres Winkes.)

on responsibility for the entire value chain, adverse health effects associated with partners, supply chains, and end-of-life disposals often go unrecognized and uncorrected. Problems only get worse when workers are engaged in the informal sector beyond the fence line of large corporations. More than half of the workers in many developing economies work in the informal sector where there are few, or no, workplace health protections (ILO, 2019). These workers contribute to the success of the global economy and bear almost all workplace health risks themselves.

Although capacity crashing usually refers to conditions outside the USA, capacity crashing and risk transfer also occur within the USA. Domestically, part-time, temporary, and gig economy workers are forced to assume risks traditionally carried by the employer. Increasingly, temporary workers and those employed by agencies supply a significant amount of labor to both large corporations and small business. These underserved workers bear a disproportionate share of the health risks in the workplace. They are also exposed to health risk associated with lack of employer-provided health insurance, long and uncertain hours, no paid sick leave, and low wages. The changing nature of work, blurring of work and home environments, and no set hours but being on 24/7 call-in status have contributed to a huge transfer of workplace health risk.

This chapter will use three examples of different industries to illustrate capacity crashing, discuss conditions that enable capacity crashing, and conclude with some suggested solutions to the problems created by capacity crashing.

8.2 EXAMPLES OF CAPACITY CRASHING AND RISK TRANSFER

8.2.1 GARMENT INDUSTRY

The garment industry is a good example of risk transfer and capacity crashing, not only in currently developing countries but throughout its history. How clothing and textiles were made changed dramatically with the advent of the industrial revolution. With increased mechanization, the textile industry significantly increased the amount of cloth produced. Textile manufacturing also required, and continues to require, large amounts of dyes and raw materials. During the 1700s and 1800s, the most commonly used raw material was cotton and the most commonly used dye was indigo. Both require intense labor, large amounts of water, and a warm to hot climate. For the USA, this meant that cotton and indigo were produced in the South primarily by slave labor.

Before the 1800s, most Americans made their clothing at home except for those wealthy enough to hire tailors or seamstresses. Ready-made garments were produced in the first half of the nineteenth century primarily for slaves. Homemade garments required time and materials that were not afforded to slaves so the ready-made clothing was a cheaper alternative. The civil war was the impetus for both the North and the South to mass-produce clothing for their soldiers, just as Rome manufactured clothing for its armies.

After the war, it became acceptable to buy ready-made clothing for women. This gave rise to many small sweatshops where mostly immigrant women sewed the garments at home for very small pay. The sweatshops were managed by contractors that worked for the emerging clothing brands. The contractors themselves were paid by the piece by the brands who then significantly marked up the price for the end consumer. In order to maximize their profits, the brands paid the contractors as little as possible. In turn, the contractors paid the actual workers a pittance and could reject any piece of clothing deemed inferior.

These conditions set the stage for the emergence of the International Ladies Garment Workers Union in 1900 and the first strike of garment workers in 1909 which resulted in a contract with higher wages for 15,000 workers. Despite an initial victory, tragedy struck 2 years later when the Triangle Shirtwaist Factory caught fire in April of 1911. The factory, like many others in New York's Garment District, occupied several upper floors of a building in Lower Manhattan where women worked in tightly packed, poorly lit spaces with little or no ventilation. The fire killed 146 people, mostly immigrant women, many of whom jumped out of windows to escape the flames. The two youngest victims were 14. The horror of this workplace fire galvanized support for garment workers and increased support for their union.

The rise and fall of the International Ladies Garment Workers Union reflects the growth and decline of the garment industry in the USA. The garment districts in New York City and Los Angeles were hubs of activity in the 1950s when most Americans bought clothing that was made in the USA. But as global markets opened up in the second half of the twentieth century, clothing companies saw the advantage of moving manufacturing to developing countries.

By the 1990s, most clothing and textile manufacturing had moved overseas. This set the stage for the development of "fast fashion" which shortened the time between design and finished product to 15 days (Schiro, 1989). Styles could change at least monthly and went out of fashion very quickly. This allowed brands to sell more clothing as consumers wanted to keep up with changing styles. In order to make room in their closets for quickly changing clothing styles, consumers needed to discard more clothing some of which had been worn once or twice, or perhaps not worn at all. Much of this clothing wound up in landfills.

In fact, the US EPA reports that in 2017 consumers discarded 8.9 million tons of clothing and footwear into landfills, with an additional 1.7 tons recycled and 2.16 tons burned in trash to energy plants. This is an astounding amount of clothing and footwear discarded in 1990 (US EPA, 2020). So not only are people and the environment exploited in the manufacturing process but disposal of discarded clothing is exacerbating problems with already overcrowded US landfills. Landfilled clothing releases toxic dyes and other chemicals, for example, long-chain per- and polyfluoroalkyl substances (PFAS), used in waterproofing, and brominated flame retardants. Discarded clothing represents an enormous amount of water and energy used to manufacture the clothing. The whole cycle of "fast fashion" is a terrible waste of resources and human capital.

The garment business has recently begun to expand in Los Angeles and, to a lesser degree, in New York. The problems that have plagued the garment industry from its inception, however, have not gone away. Most factories in LA are small with between 20 and 50 workers. Workers are paid by the piece, and overtime is mandatory. The typical length of operation for these small factories is 13 months. By declaring bankruptcy and reopening under a different name, they make it difficult for workers to claim back wages and for regulatory agencies to bring them to justice (Verite, 2015). Unfortunately, these developments are an example of domestic capacity crashing.

8.2.2 ELECTRONICS INDUSTRY

Although electric power was utilized in the nineteenth century, the electronics industry did not emerge until the twentieth century the first half of which depended on the vacuum tube. The invention of the metal–oxide–semiconductor field-effect transistor in 1959 started the semiconductor industry and provided the basis for the consumer electronics industry as we know it today. In the 1960s and 1970s, most consumer electronics manufacturing was done in the USA and Japan. But by the 1980s, the electronics industry, just like the fashion industry, moved to developing countries.

Consumer electronics includes a wide variety of things for personal use such as televisions, phones, laptops, desktops, tablets, and a host of others. In 2015, the global consumer electronics industry generated $283 billion in sales. Consumer products are projected to generate $2.9 trillion dollars in sales globally by 2020 (PR Newswire, 2017).

The occupational hazards associated with the electronics industry include solvents, metals, noise, and musculoskeletal risks. Two of the solvents considered most dangerous are benzene and n-hexane. Many factories require people to work 12-hour days, often 7 days a week. Workstations are standardized with no adjustability. Clean rooms have low humidity and restricted ventilation. Many of the workers are young women who move to urban areas from rural villages. Reports have surfaced of miscarriages, reduced fertility, and cancer (Kim et al., 2014). Most of the products were at first sold in developed countries, but now cell phones are purchased throughout the world and consumer electronics are increasingly bought by Chinese and other emerging economies.

The disposal of consumer electronics at the end of their useful life is perhaps more hazardous than the manufacture of such items (LeBlanc, 2019). Many items are built to be obsolete within a few years. Even without built-in obsolescence, advances in electronics have occurred at a rapid pace requiring frequent replacement of outdated systems. For whatever reasons, the amount of consumer electronics that is discarded is astounding. The World Economic Forum estimates that there were 47.5 million tons of e-waste in 2018 (Ryder and Houlin, 2019). The amount of electronic waste each year is expected to keep rising because of increasing demand and falling prices for consumer electronics throughout the world.

Until 2018, when China banned the import of waste including e-waste, about 70% of global e-waste was shipped to China (Semuels, 2019). Even before China's ban on accepting waste, significant amounts of electronic waste were being shipped to other Asian, African, and South American countries (Neitzel and Sayler, 2018). Unlike the manufacture of consumer electronics that is done in factories typically run by large international companies, recycling of electronic waste is done in the informal sector, which is not regulated and has few, if any, safety and health protections.

Electronic waste includes precious metals such as gold, silver, and platinum. These metals are typically integrated into the circuit board and are very difficult to extract. One way of extracting the gold and silver is to use mercury as an amalgam.

The metals "dissolve" in the mercury. The mercury can then be distilled off to leave the purified gold or silver. When this is done under controlled environments, such as a laboratory or well-designed recycling facility, the mercury vapor can be captured and recycled safely. Most of the e-waste recycling is done, however, in the informal sector where people heat the mercury amalgam over an open fire. In the process of extracting the precious metals, the mercury fumes are dispersed into their homes and neighborhoods (Puckett et al., 2002). Exposure to mercury is one of the tragic health consequences of processing electronic waste in the informal sector.

8.2.3 GOLD MINING

Mining is a dangerous occupation under almost any circumstances. Large gold mining companies have modern equipment and typically pay attention to safety standards and provide protective equipment to employees when required. Gold mining produces about 74% of yearly gold production with 15%–20% of that originating in the informal sector mined by artisanal miners. Artisanal gold mining is done by individuals without any safety or protective equipment and can be extremely hazardous. The rest of yearly gold production comes from gold recycled primarily by the informal sector.

Before the gold is sold, it has to be refined. About 70% of the world's gold is refined in Switzerland with lesser amounts being refined in Dubai and Shanghai, among other places. Gold from mining companies, artisanal miners, and recycling is combined before it reaches the refiners. There is no way to know the exact origin of gold sold by the refiners.

The gold refiners sell their gold to various companies including electronics companies that use the gold in their products. Even though there is a very small amount of gold in electronic equipment, the large number of electronics items sold each year accounts for a significant quantity of gold. Electronic companies used almost 7% of the gold produced in 2014 (Schipper and de Haan, 2015).

8.3 CONDITIONS THAT ENABLE CAPACITY CRASHING

The personal and financial cost of capacity crashing is enormous. The ILO estimates that more than 4% of global GDP is lost due to worker illness, injury, and death (ILO, 2019). These losses are often more prevalent in developing countries because they lack the capacity to deal with the hazards and stressors associated with increasing industrial activities.

Conditions that enable capacity crashing include, but are not limited to, the following:

- Although most developing countries have health, safety, and labor regulations that govern formal workplaces like factories, they often do not have the government structure and trained personnel to effectively enforce those regulations.
- Higher percentages of the population of developing countries are involved in the more hazardous occupations such as construction, mining, and agriculture without appropriate protections.

- Many workers in developing countries work in the informal sector which operates outside any legal and/or social protections that might be available to those working in the formal sectors. According to the ILO, informal work is precarious, unhealthy, and unsafe (ILO, 2016, 2019; Hussmanns, 2004).
- Strategies and systems to help and encourage employers and workers implement safe, healthy working conditions are often not as strong as they should be or as the developing country itself would like them to be.
- There may be little or no social or legal recourse for workers who are not treated fairly.
- There is often no sick leave, and worker compensation programs are weak or nonexistent.
- Education and training is needed, even in developed countries, to increase the demand for better working conditions.
- When abject poverty and hunger are the alternatives to working in a severely risky job, the occupational risk may be perceived as more acceptable. If the capacity to provide a safe and healthy workplace with a fair wage is lacking, it creates an environment that fosters exploitation and enables enterprises, whether headquartered in-country or out of the country, to overwhelm whatever health and safety capacity does exist.

All of these conditions may not be present in every situation, but one of any combination of these conditions indicates an environment supportive of capacity crashing.

Supply chains, the informal sector, and the COVID-19 pandemic are three environments where many of these conditions can be observed and where much of the occupational risk is transferred disproportionately to the workers.

8.3.1 Supply Chains

Workers in garment factories and their supply chains are often far from their home villages and families and sometimes dependent on their employers for housing. They are typically paid by the piece. Worker compensation and sick leave are usually not available. These conditions set the stage for capacity crashing.

Global fashion supply chains are often very complicated and opaque. For example, fashion brands contract with many factories, each of which makes garments for more than one brand. In addition, many of those factories have supply chains that consist of smaller industries that manufacture thread, buttons, and clothing pieces. These items are then used in the larger clothing factories to assemble the finished piece of clothing. Factories typically contract with multiple brands, and individual brands may use many different factories. Even when consumers want to support worker and environmental health by purchasing ethically and sustainably made clothing, it is very difficult to know where any one individual piece of clothing was made. This makes it almost impossible for consumers to choose to buy clothing made in the factories that have safe working conditions, fair wages, and environmental responsibility.

Similarly, the supply chain for the gold industry includes large mining companies, artisanal miners, and e-waste recyclers. Artisanal miners and e-waste recyclers account for at least a third of the gold entering the supply chain each year. Mining is hazardous under controlled conditions. Artisanal mining is significantly more dangerous, as is gold recovery from electronic waste. Like the garment industry, there is no way to identify where each specific ounce or quantity of refined gold originated. This precludes purchasing ethically produced gold by the end user even when consumers want to make responsible purchases.

8.3.2 INFORMAL SECTOR

People in the informal sector often lack knowledge about the adverse effects of what they are doing. Even with knowledge of detrimental health and environmental effects, they do not have the means to protect themselves, either with personal protective equipment and/or with environmental controls. Furthermore, informal sectors in most countries are not regulated by any governmental oversight. The informal sector is a prime example of a set of conditions that facilitates capacity crashing.

When most of the work is performed by the informal sector, the workers are almost entirely outside any legal or social protections. No one is responsible for the workplace in the informal sector. There are no sick days or worker compensation. Wages are precarious and at the whim of whomever buys materials or services from informal workers. Most informal workers are continually confronted with food, housing, and other basic need insecurities. Clean water is often unavailable, and health care is beyond the reach of most.

8.3.3 SARS-CoV-19 PANDEMIC

The current pandemic illuminates another aspect of capacity crashing which is the transfer of economic risk in addition to occupational and environmental risk. The pandemic has resulted in a sharp decline in clothing purchases in developed countries. Clothing factories had, however, already manufactured the clothing which had been ordered by the clothing brands before the pandemic struck. Many brands refused to pay for the completed orders when they were unable to sell the clothing. The global decline in clothing purchasing and the refusal of many brands to pay for completed orders transferred economic risk to already vulnerable workers in developing countries. In many cases, workers lost not only their jobs but also the housing and food that came along with those jobs (The Economist, 2020).

8.4 SOLUTIONS

Interventions particularly effective against capacity crashing or risk transfer include building and strengthening occupational and environmental health systems within the country itself, empowering worker organizations, developing verification and/or advocacy programs, education and training, and technology. Education programs should strive to reach not only workers but managers, owners, policymakers, legislators, and the consumers themselves. Technology may also be able to help by designing machines and equipment whose components can be easily separated and recycled when they become obsolete or are no longer useful. Blockchain technology can make the origin of clothing, gold, and many other consumer products crystal clear to the

TABLE 8.1

List of Organizations Available to Assist with Capacity Crashing Interventions

Organization	Website
Adalinda	https://www.adalindafashion.com/
Developing World Outreach Initiative	https://aiha-ncs.org/content.php?page= Developing_World_Outreach_Initiative
Fair Trade Certified	https://www.fairtradecertified.org/
GoodWeave	https://www.adalindafashion.com/
Maquiladora Health and Safety Support Network	http://mhssn.igc.org/
OK International	https://www.okinternational.com/
Occupational Hygiene Training Association (OHTA)	http://ohlearning.com/
Remake	https://www.adalindafashion.com/
Workplace Health Without Borders (WHWB)	http://www.whwb.org/
Workplace Health Without Borders-USA (WHWB-US)	https://whwb-usa.org/
World Bank	https://www.worldbank.org/
World Health Organization	https://www.who.int/

consumer. Organizations available to assist with capacity crashing interventions are shown in Table 8.1.

8.4.1 Capacity Building

Strengthening the occupational health and safety systems within the developing country itself is important for many reasons. Each country has its own laws and regulations and has its own approach to workplace health and safety. Governments of many developing countries recognize the need to protect their workers by developing informed and well-trained occupational health professionals and factory inspectors. Occupational health professionals in developing countries often include physicians, scientists, engineers, and environmental specialists who, although well-educated, may not have industrial expertise. The applied knowledge and skills that industrial-/occupational hygienists in the developed world have honed during the past 100 years are extremely valuable to the occupational health professionals in developing countries. Global sharing of occupational health and safety enables developing countries to learn from the experience and past mistakes of already developed countries and provides a forum for a discussion of emerging threats to workplace health and safety.

Both the World Health Organization and the World Bank promote capacity building in developing countries. The World Bank will often require the implementation of occupational and environmental improvements as a condition of its loans. Non-government organizations, such as OK International, the Developing World Outreach Initiative, and Workplace Health Without Borders and its branches, along with many others, are also involved in capacity building to give developing countries the ability to obtain, improve, and retain the skills, knowledge, tools, equipment, and other resources needed to protect worker health within their own boundaries.

8.4.2 EMPOWERING WORKER ORGANIZATIONS

Empowering worker organizations is key to workplace health. Workers know their jobs better than anyone else and can tell anyone interested in listening, details about the most difficult and dangerous parts of their jobs. Although they may not have the insights and skills to fix them, workers can often recognize hazards that should be controlled or eliminated. Workers need to be viewed as integral parts of any enterprise at the center of successful businesses. Although workers need safe working conditions, fair pay, and respect, workplace conditions are tied to forces outside their control and the control of their organization. These forces include global and country-wide market conditions, economic policies and practices, political activities, and natural disasters. This interconnectedness with external factors requires that workplace improvements involve addressing issues that are more encompassing than just the workplace. Labor advocates and labor unions understand this. So do managers and employers, which is part of the reason that there is tension between union and management in some situations.

Factory workers in many developing countries have moved to the city, or wherever the factory is located, to work. In many cases, they have left their family and social support systems behind. The disruption of traditional social networks by global industrialization makes many factory workers more vulnerable with fewer resources to fight exploitation. Trade organizations, labor unions, and religious and/or social support groups can help to fill the loss of more traditional support groups.

The ILO and the WHO (LaDou et al., 2018), long advocating for safe workplaces and worker rights, and the Maquiladora Health and Safety Support Network, empowering worker groups since 1993, are three groups dedicated to improving the workplace through worker unions and organizations.

8.4.3 VERIFICATION/ADVOCACY

A variety of schemes have been developed to verify workplace conditions and address such issues as child labor and fair pay, as well as health and environmental impacts. Typically, there is a third party, often a non-governmental organization makes the evaluation regarding the issue. For example, Fair Trade Certified, established in 1998, has developed industry-specific standards for apparel, home goods, seafood, and agricultural products. They have a process to handle complaints and update their standards every 5 years. Their goal is to foster environmental stewardship, income sustainability, and individual and community well-being. Other groups, such as GoodWeave, focus on eliminating child labor in global supply chains. GoodWeave was established in 1993, and its founder, Kailash Satyarthi, shared a Nobel Prize in 2014 with Malala Yousefzai. These are just two examples of the many groups who are using a verification approach to decrease the adverse effects of capacity crashing and risk transfer.

Fashion designers are also addressing transparency and accountability in the fashion industry. Two examples of this work are Remake and Adalinda. Remake is a group of young sustainability advocates for occupational and environmental health in the fashion industry. The goal of Remake is to make fashion a force for good; to educate, inspire, and uplift community voices in support of a sustainable future; and to help the consumer "see" and connect with the people from all over the world who make clothing they wear. Adalinda is a sustainable fashion company committed to social justice and raising awareness of the inequities in the fashion industry. Both offer lists of brands that meet specific criteria of sustainability and worker health and justice.

8.4.4 EDUCATION

Worker education and training is important to foster healthy workplaces, but the level and type of education must be tailored to the needs of the workers (ILO, 2017). The informal sector may be bigger than the formal sector in many developing countries. Informal sector workers often have no boss or set salary, and often rely on middle-men for compensation. If informal workers get hurt or sick, they are unable to work and do not get paid. Frequently, they are not aware of the hazards associated with their work and, when or if aware, have no means to protect themselves. The first step to protect informal workers is to educate them about the hazards they face. For example, a professor and her students in Brazil are working with garbage pickers to help them understand the hazards they face every day and especially during the coronavirus pandemic (Cruvinel, 2018).

Education of factory workers may look more like worker education in developed countries. Providing people with the information, and giving them the tools, they need to work safely increases their ability to do a better job. A more structured work environment also provides the opportunity to educate managers on how a culture of safety and health can not only improve quality and productivity, but also reduce accidents and days away from work.

The Occupational Training Association grew out of a recognized need to educate managers and workers in developing countries so they could understand how to operate facilities run by multinational corporations, specifically British Petroleum. This initiative has grown into a global enterprise that provides a menu of courses in various subspecialties of occupational hygiene which are designed for a variety of industries. Free course syllabi, supporting material, and slide decks are available on their website (www.ohlearning.org). If students want a certificate, however, they need to take and pass an examination offered through the British Occupational Hygiene Society.

Consumers and the general public everywhere need to be educated about occupational health and safety as well. They need to learn to "see" the workers who make the products and services that they buy. Most consumers do not want to wear clothes or gold that injured someone, or use electronics that made someone sick, but most of the time shoppers think primarily about the cost of the item they are purchasing. The workers who make the clothes we wear, who harvest the food we eat, who make the electronics we buy, and who supply the gold we purchase are invisible to most consumers. A pilot study done as part of the Visiting Partners Program at the University of Michigan Center for Occupational Health and Safety Engineering indicated that more than half of 100 consumers queried would pay more for clothing if they were sure that the money would actually go to the workers who make clothing (O'Reilly,

2018). Even if half of the consumers were motivated to care about workers, they could make a huge difference in occupational health.

8.4.5 TECHNOLOGY

Making supply chains transparent would give consumers a way to assess workplace conditions. The opaqueness of supply chains allows companies to claim ignorance about the working conditions where their clothing is made. This makes it difficult, if not impossible, to know where a single piece of finished clothing was made. It is interesting that the International Ladies Garment Workers Union developed a system of letters and numbers in the 1950s to specifically identify the factory of origin when clothing was still mostly made in the USA. The system was never widely adopted, and within the next few decades, clothing manufacturing moved offshore to developing countries where it remains today. Application of blockchain technology to the clothing supply chain, however, could significantly increase supply chain transparency (UL EHS Sustainability, 2017). Blockchain technology is already being used by some corporations, such as Walmart and seafood companies. It could be sued in a variety of industries that employ supply chains.

Better design of electronics products before manufacturing could facilitate the recovery of gold and other precious metals at the end of the products' useful life (Cho, 2018). Designing a process to extract precious metals under controlled conditions would also reduce worker exposure as well as the amount of "new" gold mined each year. The case for better original design and the circular economy is made by the fact that there is more gold in a ton of discarded phones than in a ton of gold ore (Ryder and Houlin, 2019).

8.5 CONCLUSIONS

In conclusion, capacity crashing describes the disproportionate transfer of workplace risk to vulnerable workers. The workers are often in developing countries but can also be low-wage workers within developed country. The lack of protections such as social nets, legal resources, sick leave, worker compensation, unions, and effective implementation of workplace health and safety regulations provides an environment that enables capacity crashing. Capacity building, empowering worker organizations, advocacy/verification, education, and appropriate application of technology are some of the approaches that can reduce capacity crashing and encourage community building and more equitable lifestyles.

REFERENCES

Cho, R. 2018. What Can We Do About the Growing E-Waste Problem? https://blogs.ei.columbia.edu/2018/08/27/growing-e-waste-problem/ (accessed May 10, 2020).
Cruvinel, V. August, 2018. Program Stop, Think and Dispose: A Multidisciplinary Approach to the Dialogue between the University, Community and Recyclable Materials Collectors. *You Tube*. Brazil: University of Brasilia. https://www.youtube.com/watch?v=0sPazpRkSNg (accessed May 24, 2018).

Hussmanns, R. 2004. Measuring the Informal Economy: From Employment in the Informal Sector to Informal Employment. Geneva: ILO. https://www.ilo.org/wcmsp5/groups/public/---dgreports/---integration/documents/publication/wcms_079142.pdf (accessed May 30, 2020).

ILO. 2016. Non-Standard Employment around the World: Understanding Challenges, Shaping Prospects. Geneva: ILO. https://www.ilo.org/wcmsp5/groups/public/---dgreports/---dcomm/---publ/documents/publication/wcms_534496.pdf (accessed May 30, 2020).

ILO. 2017. Improving Safety and health in Global Supply Chains, https://www.ilo.org/safework/projects/WCMS_554169/lang--en/index.htm (accessed May 5, 2020).

ILO. 2019. Safety and Health at the Heart of the Future of Work: Building on 100 Years of Experience. Geneva: ILO. https://www.ilo.org/wcmsp5/groups/public/---dgreports/---dcomm/documents/publication/wcms_686645.pdf (accessed May 29, 2020).

Kim, M.H., H. Kim and D. Paek. 2014. The health impacts of semiconductor production: an epidemiologic review. *Int J Occup Environ Health* 20: 95–114. https://www.ncbi.nlm.nih.gov/pmc/articles/PMC4090871/pdf/oeh-20-02-095.pdf (accessed May 11, 2020).

LaDou, J., L. London and A. Watterson. 2018. Occupational health: a world of false promises. *Environmental Health* 17: 81. https://doi.org/10.1186/s12940-018-0422-x (accessed May 10, 2020).

LeBlanc, R. 2019. E-Waste and the Importance of Electronic Recycling. https://www.thebalancesmb.com/e-waste-and-the-importance-of-electronics-recycling-2877783 (accessed May 10, 2020)

Neitzel, R.L. and S.K. Sayler. 2018. Global Risk Transfer Via Informal Electronic Waste Recycling. New Orleans, LA: Society for Risk Analysis, December 3–5.

O'Reilly, M. 2018. Transferred Risk in the Garment Industry and the Changing Role of Consumers. New Orleans, LA: Society for Risk Analysis, December 3–5.

PR Newswire. 2017. Persistence Market Research. Global Consumer Electronics Market to Reach US$ 2.9 Trillion by 2020- Persistence Market Research. https://www.prnewswire.com/news-releases/global-consumer-electronics-market-to-reach-us-29-trillion-by-2020---persistence-market-research-609486755.html (accessed May 10, 2020).

Puckett, J., L. Byster, S. Westervelt, R, Gutierrez, S, Davis, A. Hussain and M. Dutta. 2002. Exporting Harm: the High-Tech Trashing of Asia. San Jose: Basel Action Network, Seattle and Silicon Valley Toxics Coalition, pp. 1–54. http://www.svtc.org/wp-content/uploads/technotrash.pdf (accessed May 5, 2020).

Ryder, G. and Z. Houlin. 2019. The World's E-Waste Is a Huge Problem. It Is Also a Golden Opportunity. *World Economic Forum*. https://www.weforum.org/agenda/2019/01/how-a-circular-approach-can-turn-e-waste-into-a-golden-opportunity/ (accessed May 11, 2020).

Schipper, I. and E. de Haan. 2015. Gold from Children's Hands. Amsterdam: Stichting Onderzoek Multinationale Ondernemingen (SOMO), pp. 1–102. https://www.somo.nl/gold-from-childrens-hands/ (accessed May 5, 2020).

Schiro, A.-M. 1989. Two New Stores That Cruise Fashion's Fast Lane. *New York Times*, p. 46.

Semuels, A. 2019. Is This the End of Recycling? *The Guardian*, March 5. https://www.theatlantic.com/technology/archive/2019/03/china-has-stopped-accepting-our-trash/584131/ (accessed May 10, 2020).

The Economist Editor. 2020. Suffering from a Stitch: Bangladesh Cannot Afford to Close Its Garment Factories. *The Economist*, May 2, p. 28. https://www.economist.com/asia/2020/04/30/bangladesh-cannot-afford-to-close-its-garment-factories (accessed July 12, 2020).

UL EHS Sustainability. 2018. Blockchain for Your Supply Chain. https://www.ulehssustainability.com/blog/supply-chain/blockchain-for-your-supply-chain/#sthash.zUOtPG5d.dpbs (accessed May 5, 2020).

U.S. EPA. 2020. Facts and Figures about Materials, Waste and Recycling. Non-Durable Goods: Product Specific Data. https://www.epa.gov/facts-and-figures-about-materials-waste-and-recycling/nondurable-goods-product-specific-data#main-content (accessed May 11, 2020).

Verite. 2015. Garment Workers in the United States: Undocumented Workers and Their Vulnerability. White Paper. https://www.verite.org/wp-content/uploads/2016/11/Garment SectorUS-WhitePaper-102215-Final.pdf (accessed May 5, 2020).

9 Challenges of Managing Occupational Health, Safety, and Environment on International Hydropower Construction Projects

Laurence Svirchev
Svirchev OHS Management

CONTENTS

9.1 INTRODUCTION

The World Bank has estimated that over 1 billion people, mainly concentrated in Asia and Africa, live without access to electricity and over 3 billion people have no access to clean cooking fuel (World Bank, 2017). The World Energy Council estimated two out of ten people in the world live without access to electricity, and that in 2030, between 733 and 885 million people would still lack access to electricity. Achieving universal electricity access requires US $36–49 billion in annual investments, which is less than 4% of the current global annual investments in electricity infrastructure (Panos et al., 2016). The lack of clean, renewable energy is a significant impediment to the developing world's ability to enhance economic development, refrigerate food and vaccines, and for children to do homework at night. Conversely, the burning of charcoal, wood, and other organic fuels such as dung kills millions every year (SE for All, 2017).

In the same world in which so many cannot refrigerate their food and cook over clean fuel, each year 2.78 million workers die from occupational accidents and work-related diseases. Of these deaths, 2.4 million are disease-related (ILO, 2019). One of the most significant ways to supply clean, renewable energy is hydroelectric power, and inevitably, some of the workers who cannot return to a home-cooked meal at the end of their workday will be hydroelectric construction workers.

Historically, workers have suffered dearly in building the sources of harnessing electric power. The Hawk's Nest tunnel project in West Virginia, USA, lasted 18 months in 1930–1931, but in that time, it was estimated that 764 workers died of acute silicosis, mainly men of African-American heritage (Stalnaker, 2006; Cherniack, 1986). The manual jackhammer drilling and blasting techniques of tunnel boring that generate massive amounts of fractionated silica still exist today, although we have a much better understanding of control methods.

Hydroelectric power through the selective harnessing of waterways is one of the ways of delivering renewable nonpolluting electric energy to lift whole nations of low socioeconomic status out of poverty while enhancing sustainable growth (Chen and Landry, 2018). A reliable and powerful electric grid is a means to power other infrastructure development projects (IDPs) such as mining, rail and road, and ports.

This chapter explores the conditions faced by managers of occupational health, safety, environment, and social conditions (HSE) on hydropower construction projects (HCPs) in the developing world. The chapter is based upon review of associated available literature, but also from the author's experience managing HSE for a Chinese international infrastructure construction company in the African Sahel (Niger, a mining IDP), and HCPs in the Western Pacific (Fiji), an island between the Indian and Pacific oceans (Indonesia), and on the South Asian landmass (Pakistan). Each of these countries possesses distinctive national cultures and languages which influence how effective HSE practices are brought about. A photograph of a large Chinese dam project is shown in Figure 9.1.

9.2   SOME CHARACTERISTICS OF MANAGING HSE

Managing HSE on an International Construction Project (ICP) has characteristics significantly different than found in fixed production facilities. The construction process involves a dynamic four-dimensional flux in which geophysical conditions and hazards change on a daily basis. For example, a forest that has stood in place for thousands of years on a steep river canyon slope is selectively felled, and in its place appears first a dirt track, and then a concrete road bound by uphill and downhill

FIGURE 9.1   The construction site of a Chinese dam project. (Photo courtesy of Putzmeister.)

slopes with drainage ditches poured into place to prevent erosion. Over the course of the construction period, the different seasons can bring about radically different weather conditions, ranging from extreme hot periods with high potential for heat-related stress and exhaustion diseases to monsoon periods with high potential for slope slip.

In spite of all attempts at rigorous planning, a well-prepared plan to cut that new road through newly opened mountain wilderness in the tropics may fail because there was a 3-day heavy rainfall and a slope collapsed. In the African Sahel, a project can go from extreme sandstorm to desert flooding in a matter of weeks. The potential for such rapid changes in environmental conditions during construction puts workers, equipment, and business integrity in jeopardy.

HSE managers therefore need not only to be versed in the fundamentals of disease prevention and safety engineering, but also have the ability to anticipate and recognize the environment-specific hazards as their risk level changes in the course of construction. They must be able to rapidly improvise work procedures in concert with the construction department without compromising the health and safety of workers.

Professionals trained in both occupational hygiene and safety are typically expected to also manage environmental issues. These issues include the prevention of land degradation, protecting endemic wildlife, flora, and fauna. The HSE manager may not have been trained in these issues and must have the flexibility and openness to become familiar with them.

9.3   ORGANIZATIONAL STRUCTURE OF AN ICP AND RELATION TO THE CONTRACTOR'S HSE DEPARTMENT

ICP is typically organized according to a contractual structure known as Engineering, Procurement, Construction, and Commissioning, shortened in construction lingo to EPC (Picha et al., 2015; Ding et al., 2018). On EPC, a private or a state power company is the owner, which hires a supervising engineer to ensure the project is completed on time and according to the design specifications accepted by the owner. The owner also hires a contractor to design and construct the works according to a controlled master schedule. Ultimately, the owner is vested with all authority on an ICP.

It is the owner who attempts to secure low interest loans from private internal sources and other-country financiers, including the World Bank (WB), its affiliate the International Finance Corporation (IFC), and the Asia Development Bank (ADB). These fiduciary organizations also require, as conditions of loan, that there be mechanisms in place to ensure social and environmental oversight of the Hydropower Construction Project (HCP). The International Labor Organization (ILO) specializes in programs to protect the rights of workers. Social oversight includes safe conditions of work, fair remuneration, work-rest cycles, and gender equality. Social oversight also includes fair payment for land purchases, respect for local customs and faith, and mechanisms for conflict resolution according to recognized international conventions and laws of the host country.

While sound in principle, the WB, IFC, ADB, and ILO guidelines are of necessity generic in nature and need to be professionally applied according to the specific social, geophysical, meteorological, and hazard conditions of every project according to laws of the host country (IFC, 2014; ILO, 2017). An example is that under ILO Article 8(1) of Convention No. 167, the principal contractor is responsible for coordinating safety activities of all parties on site (ILO, 1988).

Developing countries may not have enough engineering expertise to build an ICP solely based on in-country resources. They therefore outsource tenders. Chinese ICP companies have become frequent winners in the tendering process due to their vast engineering and design expertise, state-backed financial muscle as part of the Belt and Road strategy, experienced workforces, and the ability to manufacture essential equipment such as turbines and generators at low cost. In other words, Chinese companies can often outbid the international competition (Chen and Landry, 2018).

In an EPC project, the contractor therefore has multiple responsibilities. One of these is to develop a program that protects the health and safety of workers, the social conditions under which they work, and to manage environmental affairs. The role of the contractor HSE manager requires weaving, hopefully with finesse, through multiple business structures to accomplish multiple HSE goals while allowing construction to proceed without problems and on schedule.

In the best of circumstances, there is mutual shared vision with the owner and supervising engineer about how to complete a complex set of construction elements spread over large areas of land. Inevitably, there will be conflicts about the conduct of business. Sometimes, these are the result of pressures from the social, environmental, and political communities outside the HCP home country, such as international

NGOs concerned with environmental and biodiversity issues. The contractor's HSE manager is typically a foreigner who needs the empathy and willingness to understand, without chauvinism, the social conditions of the project country, some of which have yet to completely heal from the colonial era. Typically, these skills come about through the ability to "stop-look-listen" before making judgment calls on OSH behavior of different nationalities and cultures.

9.4   FUNCTIONS OF THE CONTRACTOR HSE DEPARTMENT

The responsibilities of the contractor's HSE functions are established in the project contract. The contractor is responsible for designing, implementing, and controlling a program designed to prevent accidents, serious injuries, occupational diseases, and fatalities. To accomplish these goals, an efficient method of driving the program from general to particular is the fundamental approach of the industrial hygiene profession: anticipation, recognition, evaluation, and control of hazards (BGC, 2020).

On EPC ICPs, the three basic contractor functions are construction, quality control, and HSE work whose activities need to be coordinated. In addition to the normal functions of accident and disease prevention, such as developing safe work procedures and training workers in them, the HSE department is typically charged with the provision of medical and first aid services.

Importantly, an ICP is a high public visibility entity and will be held to a higher standard of public health and environmental responsibilities than may exist in local remote and underdeveloped communities. The responsibility to prevent environmental degradation therefore assumes an outsized proportion of an HSE department's time and activity. Owners are acutely sensitive to any local public and international criticism of environmental degradation or harm to wildlife. The HSE department typically has responsibilities for camp and living quarters sanitation, pest control, and instrumental environmental sampling and monitoring. The monitoring includes ensuring that human waste does not enter waterways and that garbage is properly disposed of.

The environmental section of an HSE department is responsible for checking, for example, settling ponds for runoff from rock crushing operations and inspecting for the existence and maintenance of designated spoil piles for forestry and tunneling operations. Environmental issues also include coordinating construction activities to ensure that disturbance of flora and fauna, especially endangered species, is minimized. Every contract will also require that as the construction comes close to the commissioning date, the environment be restored as close as possible to its original native state.

Security services are required for protecting the blasting magazine, preventing pilfering, guarding equipment left overnight in remote areas, controlling political protests to ensure no one is hurt, and ensuring the business integrity of the ICP is not compromised by accidents and fatalities. HSE departments are therefore involved in crossover activities coordinated with the owner for site security, local police forces, and in some cases even the armed forces.

Case Study

One ICP had the distinction of having endangered species on the project. One species was tiger, which could endanger human lives. There were also orangutans and monkeys. Given that a sense of danger from tigers automatically overcomes the human sense of inquisitiveness, workers had no qualms about stopping work when tigers were thought to be near a work area.

The orangutans and monkeys, in contrast, are rather cute, relatively benign, and attract sincere curiosity when they are observed close to work areas. Workers really don't mind stopping work to observe them from the required distance. In the local culture, the orangutans are considered as sacred and their lives may not be disturbed. The orangutans typically ignore human activity and take their own time moving in and out of work areas.

Sightings of tiger were extremely rare, since they are stealthy animals. Sightings of orangutans were relatively frequent. In both cases, there are specific stop-work and wait procedures to ensure neither the animals nor workers are directly endangered by construction activities. There are also procedures, strict in nature, to radically limit impact on wildlife. These include arbor-bridges. If necessary, a specialist from a local environmental NGO is called to ensure there is no danger to any wildlife species.

9.5   A GREAT COMING TOGETHER OF PERSONNEL ON AN ICP

Contractors are usually obliged by international convention and conditions of loans to hire the workforce from the host country. The exception to the general rule is in cases in which the host country does not have skilled workers to meet the construction needs. The contractor mobilizes a professional design company, subcontractors, and skilled workers from its own country to manage and supervise the construction activities of workers from the contractor's home country. ICP typically takes place in countries with large rural and remote areas characterized by low socioeconomic, health, and educational conditions. Since the economy of these areas is frequently based on agriculture and forestry operations, the local laboring workforce tends to be unskilled with sparse experience working in the high intensity, high hazard labor conditions (ILO, 2015).

While many developing countries have trade schools for skilled workers such as mobile crane operations and welders, the local manual laboring workforce on an ICP "learns by doing," that is, emulating the work practices of the contractor's employees. This can be acceptable practice if the contractor's employees are working according to accepted safe work procedures.

In the conditions of an ICP, there is a great coming together of multiple nationalities and cultures, each with their own work habits, expectations, cultures, and approaches to government regulations. An OHS professional trained in Canada, France, or China will necessarily have different appreciations of how to strategically manage HSE depending on their background and experiences in their home country. Managing HSE requires not just the technical mastery of industrial hygiene and safety and environmental engineering, but the ability to adapt the technical skills into alignment with the cultural characteristics of multiple nationalities. The coming together of nationalities into a rural, low socioeconomic terrain with little experience with foreigners is usually an enlightening experience, but it can also be a breeding ground for national chauvinism to the point of violence when a small number of foreigners do not treat local people with respect.

Reducing the potential for violence is no trivial matter for an HSE manager who arrives in a new country having meager knowledge of local culture. Some ICPs are built in areas that have seen recent civil wars based on ethnic and religious conflict. This means many workers of the project country are trained in combat and insurgent techniques and may not take lightly to actual or perceived racism or superiority on the part of foreigners.

Violence has no place in the workplace. The adept HSE manager must be prepared to recognize the signs and symptoms of racism or national chauvinism and stop it in its tracks, typically through the mechanism of education. It may be necessary to send racist foreign workers back to their home country, both for their own protection and to protect the business integrity of the project.

In most circumstances, however, the coming together of nationalities on a worksite has positive benefits. As an example: how an HSE department of a Chinese contractor promoted cultural respect, the following situation from the Sahel of Africa is offered.

Case Study

Professional drivers had been operating dilapidated trucks for many years until the ICP came to their area of Niger. These trucks had weak suspension mechanisms, poor brakes, and lacked seat belts. A brand-new fleet of dump trucks had been imported from China and the HSE team analyzed that the first intuitive action of drivers would be to test the capacities of the new trucks and speed while driving unobserved along the flat Sahel.

The HSE department held safety meetings to coach drivers in a two-point message: do not speed in their new trucks and to wear seat belts. The project language was English, yet the contractor HSE team insisted that meetings were held in a combination of English, French, Hausa, and Tamahaq languages and repeated the same messaging multiple times. Feedback from workers indicated they were pleased with the results. The contractor's approach was in radical contrast to the owner's safety meetings which were conducted exclusively in their European language. The owner's home country was also the colonialist owner of the country before independence.

9.6 OHS REGULATIONS

Within the developing world, there exists a broad range of regulations that govern HSE. Some countries have advanced occupational health and safety regulations. Others have a set of regulations adopted from the more developed former colonial powers. Some poor countries of Africa have sparse regulatory regimes. Examples are presented in the following paragraphs.

In 2003, Indonesia instituted a requirement that every enterprise with more than 100 employees must have an Occupational Safety and Health Management System with the objective of preventing occupational accidents and diseases (ILO, 2012). The plan must be based on hazard identification, risk assessment, and management. It must consider, among other requirements, control measures, design and engineering, working procedures, and purchasing/procurement of goods and services. Indonesia has no regulations specifically dedicated to construction processes.

Fiji's OHS regulations rest upon a 1996 law, the Health and Safety at Work Act 1996 (Fiji, 1996). Its system of standards and codes of practice are not home-grown, but refer to Australia, New Zealand, British, European, and American standards. For the purpose of compensation of occupational diseases, Schedule 1 takes into account diseases caused by chemical and physical agents, parasitical diseases, among others. One of Fiji's principal exports is sugarcane, and Schedule 1 includes diseases of target organs such as bronchopulmonary diseases caused by dust of sugarcane (bagassosis). Fiji has no regulations specifically dedicated to construction processes.

The poorest countries of Africa have almost sparse governmental capacity to enforce safety and health regulations and comprehensive regulations to protect workers in all sectors of the economy may not even substantially exist. Niger, for example, is the second least developed country in the world, its economy centering on subsistence crops and livestock which generate about 40% of the GDP (CIA, 2020). The country has substantial reserves of unexploited uranium, gold, and coal deposits, but these are poorly developed (Chen and Landry, 2018). The ILO website shows that there are recent regulations concerning health and safety committees and radiation/silica exposures but there are few other regulations (ILO, 2020). The country has no regulations specific to construction. During the time the author spent on an ICP in Niger, there was not a single government HSE inspection.

9.7 THE PHYSICAL CONDITIONS OF THE TERRAIN DETERMINE MULTIPLE HIGH-RISK ICP HAZARDS

ICPs take a long time to plan, develop, construct, test, and commission the project. The preparation period can be a decade-long process and the construction period 5 years or longer, depending on financial and geopolitical considerations that can change over time. Initial investigations include identification of the terrain and features such as geology, bedrock stability, fault lines, and environmental features of land flora and fauna, river flow forecasting, and weather patterns.

For example, six HCPs in Nepal for the period 2007–2025 had construction schedules of 5 years each (Bonaventura, 2015). The medium-size HCP the author has worked on has had a construction phase period of 5–8 years. The 14,800 MW Three Gorges Project in Hubei, China, began construction in 1994, and the physical dam was mostly completed in 2006 (Glieck, 2008). *Force majeure* such as war, strikes, bad weather, earthquakes, landsliding, and land acquisition can interrupt or delay construction schedules (Batool and Abbas, 2017). Large dams, no matter in what part of the world they are constructed, often take significantly longer than planners forecast. In this regard, North America with a 27% mean schedule overrun is the best performer (Ansar et al., 2014).

A 5–8-year construction phase has direct bearing on host countries with regard to chemical and physical exposures which can result in the recognition and compensation of occupational diseases. Occupational diseases with lengthy latent periods may be diagnosed long after the contractor has departed from the local area. Consequently, the disease may never be tied to the actual exposures that occurred even decades before. This concept will be explored later in the chapter.

The hazard and risk analysis of an HCP starts with a four-dimensional analysis: the three-dimensional nature of the terrain and weather, and their changes over time. HCP for hydropower involves a process of transformation of narrow and long swaths of essentially virgin land and forest, the changing of upstream, downstream, collateral waterways, and the water table. What is temporarily or permanently altered is replaced during the construction phase with tunnels, concrete and packed earth dam structures, powerhouse buildings, and new roads. The land and waterways then require pre-commissioning remediation in the form of tree and foliage replanting, creating conditions for the return of wild- and aquatic-life as much as possible to the original state, and decommissioning of temporary roads.

Each of these parameters has bearing on the health and safety of workers. Plans for sudden shifts in weather must be developed and implemented according to four-dimensional analysis by taking into account the nature of the terrain, the change of seasons, and sudden changes in weather. Sandstorms, for example, are one of the serious seasonal hazards that workers face in the Sahel of Africa. A primordial measure that must be taken in developing a site is the provision of shelter in case of sandstorm. Sandstorms give about a 20-minute warning, first as a thin line on the horizon combined with slight increase in wind pressure with light pinpricks of sand on the skin. Desert people know these conditions well, but foreign workers who have just arrived will have had no experience.

Case Study

Twenty-five foreign workers had just started to survey and level a worksite for a mineral processing building when sandstorm warning signs were observed. Shelter for these conditions had not yet been installed. The only possible shelter was one pickup truck used to ferry workers. An emergency mobile phone call was placed and the HSE department pickup truck, the only vehicle in the area, arrived just as the sandstorm hit. Twenty workers filled the cabs of the two pickup trucks until the sandstorm passed.

Mountain weather is unpredictable. There can be ferocious rain in one location, while one kilometer away, the sun is shining. HCP dams are typically built in river canyons due to increased flow rates, to reduce the span of the structure, and capture as much reservoir water as possible. During the construction phase, sudden and unseen upstream river surge flooding can engulf workers. In addition, when ferocious rain conditions are combined with denudation of steep slopes due to logging operations, there are no longer root systems in place to stabilize soil. Construction methods to remediate the hazard consist of shotcreting slopes, drainage ditches, and conduits.

Case Study

On one ICP located near the equator, a temporary drilling rig was set up next to an unfinished road on the downhill side a steep slope above a river. The rainy season had already started. The purpose of the vertical drilling was to find a suitable site for a particular type of tunnel known as an "adduct." The crew that set up the rig platform did not adequately consider the stability of the uphill slope, even though portions of that slope had already been logged out for road construction and drainage ditches were not in place. Neither the crew nor its management contacted the HSE department to survey the placement of the rig with regard to hazard and risk.

Overnight, a flash storm with extreme wind force saturated the soil on the slope above the rig, the soil slipped, and the drilling rig was washed about 75 m into the river below, never to be recovered. Fortunately, the incident happened overnight and there were no casualties. Water-saturated soil that is ready to slip is not easily observable. If the soil saturated by the overnight rain had slipped 4 hours later during working hours, five lives would have been lost and recovery of the bodies may have been impossible.

The lesson from these two examples is that the geophysical, seasonal, and weather hazards must be anticipated and recognized before they can be controlled. The successful management of HSE on ICP cannot be reduced to the simple equation of following local country or even international regulations. An HSE department's work must have the flexibility to work closely with other disciplines on site to ensure the protection and safety of workers and the continued business integrity of a project.

9.8   HAZARDS IN RELATION TO THE SOCIAL COMPOSITION OF THE WORKFORCES

HCP is a high-risk activity, and the environmental and geological conditions form important physical bases for serious injuries and fatalities. Hazards include blasting, falls from extreme height and into swift-moving water, mobile equipment turnover on steep slopes, road accidents, terrain, confined spaces, tunnel, and structure failures. Other hazards include moving machinery such as rock crushers and vibrating screens, electricity, welding, fires and explosions, and noise from machinery, and air compressors.

An HCP of 40 MW may have a total of about 500 workers pass in the construction phase, and a project of greater than 4,300 MW will have about 8,000 workers. The workforce typically consists of local-region unskilled laborers, most of whom will have had no previous experience in large-scale industrial and construction work. Skilled and semi-skilled workers will be hired from all around the host country. The skills set of these workers will include blasting specialists, drivers, mobile equipment and crane operators, welders, various grades of mechanics, electricians, penstock pipe rollers, carpenters, and painters.

In projects the author has worked on, the contractor has been a Chinese international construction company specializing in water projects. Expatriate workers, in contrast to workers of the home country, travel from project to project and spend years in host countries, sending their wages home for their families and in some cases only returning home for holidays once a year. The range of workers includes senior construction and mechanical-electrical equipment installation managers, site supervisors, and skilled workers for specialized equipment such as blasting equipment, multi-head tunnel boring machines, turbine, and electrical equipment installation.

HCPs are not like point-central manufacturing facilities. Hydropower construction takes place over long distances involving multiple staging points. Depending on the design of the project, there will be sub-sites stretched along the river for housing and feeding workers. Local workers may be commuting, but the skilled trades staff are either housed with local communities, or require on-site domestic facilities. The international workforce includes visiting and supervising consultants and managers, and they too require housing and canteen facilities in semi-permanent micro-communities. All on-site camp facilities require sanitary food preparation, eating, toilet, facilities, and potable water.

The HSE manager needs to have a good understanding of community and public health to ensure that food, water, insect, and sexually transmitted diseases do not cripple workers' health and compromise the business integrity of a project in the eyes of the public and government of the host country.

9.9   WHAT THE EMPLOYER AND THE SUPERVISING ENGINEER SEE

It is a given that many ICPs are high hazard enterprises. Most outsiders to the construction industry, and even experienced engineers, can eyeball a project and presume that falls from height represent the most serious high-risk pre-conditions for injury and death on a construction site.

In the experience of the author over the course of 9 years with Chinese contractors on three different ICPs involving extreme heights, there have been no fatalities related to falls from height, and only one serious injury as the result of a fall from height. The majority of serious injuries and fatalities resulted from other factors: loss of control of dump trucks, concrete tankers, and mobile equipment on mountain roads and extraordinarily, a dump truck rollover on a flat, gently curved desert road. All of the incidents had driver error as cause, and none of these incidents involved excessive speed. The author does not claim these findings have any statistical significance, but what they point to is that not all hazards are equal in risk, that risk will change from project to project.

On an HCP in the West Pacific, there were no serious injuries or fatalities over a 3-year period. Investigation by the HSE department did determine, however, that the principal cause of lost time was due to diseases related to weather and personal hygiene issues. Appearances should not be the driving force in establishing a prevention strategy, but real-life experience should be.

Construction engineers are of course aware, by virtue of training and experience, of the range of physical hazards. They also understand the contractual requirements for safety inspection of the contractor's work. In the experience of the author, however, employers, supervising engineers, and contractor engineers are almost never trained in industrial hygiene and the health sciences. Consequently, they typically hire HSE representatives trained only to the safety technician level and with little training in the health sciences. Even though the owner and supervising engineer have departments named HSE, the first letter in the acronym is not fully taken account of since the departments are not staffed by professionals who understand chemical, physical, and biological exposures that are precursors to acute and chronic occupational diseases.

HSE technicians are usually competent at recognizing activities that do not meet generic safety standards. They may be limited in their ability to prevent accidents and fatalities because they are not well trained in the last three conditions of the industrial hygienists' formula, "recognition, anticipation, evaluation, and control of occupational hazards." In practice, the HSE staffs of both the employer and the supervising engineer have a hard time differentiating between high-risk and low-risk activities, because they concentrate on the hazards but are not comfortable judging risk. All hazards therefore become equal. These conditions limit their utility in systematically improving safety management systems as opposed to rectifying the most serious hazards in order of priority, on a one-by-one daily basis.

9.10   COMMAND AND CONTROL MANAGEMENT STYLE

The ultimate authority on an ICP belongs to the owner of a project as guided by conditions of contract. This power can often result in owner representatives unduly using their authority and express it using the "Command and Control" (CnC) method of HSE management. CnC is based on the authority to compel a contractor to immediately correct all hazardous conditions without considering the imminence of the hazard, severity of the hazard, the present or future level of risk, and the relation of a particular hazard to other hazardous conditions. The following example illustrates how the CnC method can result in treating all risks as equal, that is, placing emphasis on low- and no-risk situations to the detriment of first resolving high-risk situations.

Case Study

The contractor HSE team was inspecting the installation of a used rock crusher in the Sahel of Africa. They determined that contract welders had entered the crushing mechanism to gouge old welds and replace crushing edges of moving machinery. The switch to activate the crusher mechanism was located next to the entry ladder; the central lockout point was in a trailer 25 m away behind a key-locked door. This presented an extremely dangerous situation since there was no way to lock out the crusher. There was no supervisor on site, and the crew insisted they could work safely, meaning they intended to continue working once the HSE team left.

While the HSE team was working to correct the situation, the supervising engineer representative called to demand that the HSE manager immediately report to a second site which did not have properly constructed stairways. This second site was an hour drive away. The HSE team said that they could not attend this second site because they were dealing with a dangerous situation and furthermore this second site was not scheduled for work for another 7 days. In spite of these extenuating circumstances, the supervising engineer commanded the contractor HSE to attend immediately under threat of discipline. The contractor HSE found an excuse for arriving 2 hours later after they had controlled the dangerous situation.

The CnC method is useful for exceptionally dangerous and emergency situations, but it is rarely helpful for the systematic day-to-day correction of worksite hazards. CnC presumes that the owner or the supervising engineer possesses superior HSE knowledge than the contractor. The CnC method of hazard management also presumes the hazards the owner or supervising engineer identifies are more important than what the contractor HSE identifies in the field. The CnC method for daily operation is built on a kind of chauvinism that breeds distrust, rather than collaboration in solving hazardous conditions. In addition, the method of CnC simply does not work in cultures in some developing countries in which softness, traditional wisdom, sympathy, and teaching by example are the marks of genuine authority and leadership.

CnC management as a norm inevitably ends up being greeted over time with "shoulder shrugs" and resistance. There is another management pathway, however, which actually involves more hard work than the issuing of rigid and unreasonable commands. This pathway involves goodwill and understanding, since each party working on an ICP has the same interests of getting the job done on time, maximum profit, and with no serious injuries, fatalities, occupational diseases, or capital equipment loss.

9.11   OCCUPATIONAL DISEASES

In 2019, the ILO published a report *Safety and Health at the Heart of the Future of Work* (ILO, 2019). The report estimated that each year there are approximately:

- 374 million non-fatal work accidents,
- 2.4 million occupational disease deaths, and
- 380,000 work accident deaths.

Based on these estimations, one obvious conclusion is that the burden of suffering and the social cost of occupational disease death vastly outweigh acute worksite accidental deaths. Furthermore, construction is a sector associated with a high number of occupational accidents and diseases. While work in construction represents between 5% and 10% of the workforce in industrialized countries and an increasing proportion in many developing countries, approximately one in six fatal accidents at work occur on a construction site (ILO, 2015).

Construction processes lead to multiple occupational diseases. Decreased lung function (FEV_1) as a result of exposure to nitrogen dioxide (NO_2) from diesel exhaust was found in tunnel concrete construction workers in Norway who did not normally use respiratory PPE (Bakke, 2004). Diseases of the lung like silicosis and lung cancer can result from tunneling operations. Dermal, acute neurotoxicity and respiratory system sensitization can result from painting, isocyanate, and solvent exposures. Physical exposures from noise can result in hearing loss; repeated lifting awkward can result in chronic musculoskeletal injuries.

Latent period is an important concept in occupational diseases. Latency is defined as "the time since first exposure (the start of the casual process) to the earliest detection of the outcome." In discussing dose, the European Lung White Book stated "High levels of dust over a long period are necessary to cause, for example, pneumoconiosis and COPD." Conversely, only a few weeks of asbestos exposure may lead to malignant mesothelioma 50 years later. This statement is reminiscent of what happened with regard to massive dose exposure to silica 90 years ago in the Hawk's Nest disaster.

Except for the very largest HCP, the construction phase will take place over period of activity in the 5–8 years range. The ICP will recruit large numbers of local unskilled workers, the majority of whom will probably go back to their previous rural life once the project is finished. While this chapter is not directed toward compensation issues, these findings demonstrate that the diseases of occupation, fatal or not, have direct bearing on the public health of general society and the costs of health care. The employers whose conditions of work created the hazards that resulted in diseases like silicosis and lung cancer will have long departed, leaving behind families and communities that may not even recognize these diseases as occupational in origin, with no ability to recoup even minimal support to those who are suffering.

9.12   THE DISEASES OF PUBLIC HEALTH IN THE OCCUPATIONAL SETTING

Case Study

The HSE manager on an ICP in Southeast Asia was having a casual conversation with the contractor's MD about traditional medicine in his country. A Chinese worker came to the clinic with an insect bite. A few minutes later, a local worker walked in with acute gastrointestinal distress. After treatment of the patients, the manager asked the MD, "What was the dominant reason for worker visits to the clinic, could the patient records be examined?"

"No need to look at the records," the MD said. "I can tell you right now that many workers take a few days off for flu, fever, GI distress, and body weakness. We don't have a big problem here with injuries. But we do have a problem with local conditions of poverty. There is only one way to deal with these issues and I am glad you asked. We need a 'Health Promotion Campaign' to keep workers healthy on and off the job. If we educate them correctly, it will improve the health of their family and their village."

As stated earlier, the contractor HSE manager is usually a foreigner and may be blinded or ignorant to many local conditions, including not making a connection between public health and occupational health. The diet of local people on this HCP was not designed for the hard labor of construction work. Their custom was to eat with their hands the cold food that they brought from home for lunch. They were not accustomed to washing their hands before eating at work. According to local MDs, the daily rain-sun cycle and changes in body temperature induced fever and flu, much in line with traditional medicine approaches to health one finds in Asia. Community sanitation was not adequate in the low socioeconomic conditions of some local villages, where running water often comes from streams, there are no flush sanitation systems for human waste, and there are community public bathing systems.

After discussion with the HSE team, particularly the in-country university-trained safety, hygiene, and social-affairs officers, the decision was made to launch a systematic "Health Promotion Campaign." The objective was handwashing upon arrival at the worksite, before lunch, and before going home. Each work team would receive training on personal hygiene from an MD or a site Nurse. The Safety Officer for each site would carry a dispenser bottle of liquid soap and make sure that everyone washed their hands according to the daily schedule. The Medical department would begin supplying a weekly report to the HSE department hygiene officer, and a process would begin to see if occupational diseases related to GI distress, flu, and fever decreased. The Health Promotion Campaign, however, was unable to bear fruit because the project was temporarily suspended due to reasons unrelated to HSE.

The lessons of the initiative were clear, however. An HSE department has to periodically examine the medical records to determine injury and disease trends and find keep an open mind and be constantly on the prowl for new ideas. Field observation is essential to the success of an HSE program, but it is not the only means to determine successes and weaknesses in the program.

9.13 CONCLUSIONS

The purpose of this chapter is to explore some of the challenges of managing the health and safety of workers as well as environment protection on international infrastructure projects, especially on hydropower projects. These projects typically have a construction phase of 5–8 years in circumstances of high hazards such as forestry, tunneling, welding, confined spaces, and working at extreme height. Hydropower and other infrastructure work frequently are carried out in the developing world in which there may or not be adequate regulatory and governmental protection of workers. Workforces are international in character and may include people of very different educational, cultural, and linguistic backgrounds, each with differing understandings of how to effectively protect the health and safety of workers.

Construction is often carried out in extreme geophysical conditions such as river canyons and steep slopes, as well extreme weather conditions. Environmental protection is a particularly sensitive issue with local people, governments, and international NGOs. In an era in which mobile phone cameras and drones can publicize alleged or real infractions almost instantaneously, owners and contractors have to proceed with their work exercising extreme diligence.

All of these conditions place a heavy burden on the HSE professional. If there is a primary lesson from the literature and anecdotal experience offered in this chapter, it is that the HSE professional needs to exercise not only the technical aspects of industrial hygiene and safety engineering, but that the professional needs to be an open-minded listener to the various cultures that form the human resources for the project. Relationships built on mutual respect with one goal in mind, the protection of workers and the preservation of environment will make successful results much easier to accomplish.

REFERENCES

Ansar, A., Flyvbjerg, B., Budzier, A., Lunn, D., Should we build more large dams? The actual costs of hydropower megaproject development. *Energy Policy* (2014), doi: 10.1016/j. enpol.2013.10.069i.

Bakke, B., Ulvestad, B., Stewart, P., Eduard, W., Cumulative exposure to dust and gases as determinants of lung function decline in tunnel construction workers, *Occupational and Environmental Medicine* (2004) Vol. 61(3), pp. 262–269. doi: 10.1136/oem.2003.008409.

Batool, A., Abbas, F., Reasons for delay in selected hydro-power projects in Khyber Pakhtunkhwa (KPK), Pakistan. *Renewable and Sustainable Energy Reviews* (2004), Vol. 73, pp. 196–204. doi: 10.1016/j.rser.2017.01.040.

BGC, IH Defined, Board for Global EHS Credentialing (2020) http://www.abih.org/content/ ih-defined. Retrieved 2020-08-09.

Bonaventura et al. Construction claim types and causes for a large-scale hydropower project in Bhutan. *Journal of Construction in Developing Countries* (2015), Vol. 20(1), pp. 49–63.

Chen, Y., Landry, D., Capturing the rains: Comparing Chinese and World Bank hydropower projects in Cameroon and pathways for South-South and North South technology transfer. *Energy Policy* (April 2018) Vol. 115, pp. 561–571. doi: 10.1016/j.enpol.2017.11.051. Accessed August 10, 2020.

Cherniack, M., *The Hawk's Nest incident: America's worst industrial disaster*, Yale University Press (1986). ISBN-13: 978-0300035223.

CIA, The World Fact Book – Niger (2020) https://www.cia.gov/the-world-factbook/countries/ niger/. Retrieved 2021-05-04.

Ding, J., Chen, C., An, X., Wang, N., Zhai, W., Jin, C., Study on added-value sharing ratio of large EPC hydropower project based on target cost contract: a perspective from China, *Sustainability* (2018) Vol. 10(10), p. 3362. https://www.mdpi.com/2071-1050/10/10/3362/ htm. Retrieved 2020-07-01.

Fiji, Health and Safety at Work Act 1996 (1996) https://laws.gov.fj/Acts/DisplayAct/454. Retrieved 2020-07-01.

Glieck, P., The World's Water 2008–2009. The Biennial Report on Freshwater Resource. Ed: Peter H Glieck, Island Press, 2008.

IFC, *Environmental and Social Management System Implementation Handbook Construction*, International Finance Corporation (June 4, 2014) https://www.ifc.org/wps/wcm/connect/ 965662f3-0cc3-4863-a962-76d31a8a979a/ESMS+Handbook+Construction-v7. pdf?MOD=AJPERES&CVID=kvLhCPJ. Retrieved 2020-07-01.

ILO, Safety and Health at the Heart of the Future of Work: Building on 100 Years of Experience, International Labor Organization, Geneva (2019). ISBN: 978-92-2-133151-3 (print). https://www.ilo.org/safework/events/safeday/WCMS_686645/lang--en/index. htm. Retrieved 2020-07-01.

ILO, General Survey on the Occupational Safety and Health Instruments Concerning the Promotional Framework, Construction, Mines and Agriculture, International Labor Organization, Geneva (September 2017) *International Labour Conference, 106th Session.* https://www.ilo.org/wcmsp5/groups/public/---ed_norm/---relconf/documents/ meetingdocument/wcms_543647.pdf. Retrieved 2020-07-01.

ILO, C167 - Safety and Health in Construction Convention, 1988 (No. 167), International Labor Organization, Geneva (1988) https://www.ilo.org/dyn/normlex/en/f?p=NORML EXPUB:12100:0::NO::P12100_ILO_CODE:C167. Retrieved 2020-08-01.

ILO, Good Practices and Challenges in Promoting Decent Work in Construction and Infrastructure Projects, International Labor Organization, Geneva (2015) https:// www.ilo.org/wcmsp5/groups/public/---ed_dialogue/---sector/documents/publication/ wcms_416378.pdf. Retrieved 2020-08-03.

ILO, Indonesia Occupational Safety and Health, International Labor Organization, Geneva (2012) https://www.ilo.org/dyn/natlex/natlex4.detail?p_lang=en&p_isn=107245&p_country=IDN&p_classification=14. Retrieved 2020-07-01.

ILO, Niger Occupational Safety and Health, International Labor Organization, Geneva (2020) https://www.ilo.org/dyn/natlex/natlex4.listResults?p_lang=en&p_country=NER&p_count=233&p_classification=14&p_classcount=14. Retrieved 2020-07-01.

Panos, E., Densing, M., Volkart, K., Access to electricity in the World Energy Council's global energy scenarios: An outlook for developing regions until 2030, *Energy Strategy Reviews* (March 2016), Vol. 9, pp. 28–49. https://www.sciencedirect.com/science/article/abs/pii/S2211467X15000450?via%3Dihub. Retrieved 2020-07-15.

Picha, J., Tomek, A., Lowitt, H., Application of EPC Contracts in International Power Projects, *Procedia Engineering* (2015) Vol. 123, pp. 397–404. https://reader.elsevier.com/reader/sd/pii/S1877705815031628?token=DF358D93534D07ECD28D3F0E99427113140BA3E76423C49730B04D66DF5EF574A86848E4E2A0477E946D08A434586321. Retrieved 2020-07-01.

SE For All, Sustainable Energy for All, Missing the Mark – Gaps and Lags in Disbursement of Development Finance for Energy Access, Sustainable Energy for All, Washington, DC (2017) https://www.seforall.org/system/files/gather-content/2017_SEforALL_FR1_0.pdf. Retrieved 2020-07-01.

Stalnaker, K., Hawk's Nest Tunnel – A forgotten tragedy in safety's history, *Professional Safety* (October, 2006) pp. 27–33. https://aeasseincludes.assp.org/professionalsafety/pastissues/051/10/021006AS.pdf. Retrieved 2020-07-01.

World Bank, Global Tracking Framework 2017- Progress toward Sustainable Energy (2017) https://www.worldbank.org/en/topic/energy/publication/global-tracking-framework-2017. Retrieved 2020-07-01.

10 An Analysis of the Impacts of National Regulation on Occupational Safety and Health

How Regulations Protect Workers and Their Employers

Ilise L. Feitshans

Nanotechnology European Scientific Institute,
ISTerrre University of Grenoble, The Work
Health and Survival Project USA and Europe

CONTENTS

10.1   INTRODUCTION: SOUND OCCUPATIONAL HEALTH PROGRAMS ARE THE GREASE FOR THE ENGINES OF COMMERCE, NOT THE FAT TO BE TRIMMED IN LEAN ECONOMIC TIMES

"The protection and promotion of the health and welfare of its citizens are considered to be one of the most important functions of the modern state" (Rosen, 1958). It is not surprising therefore that, throughout recent history, precautionary principles of science have been embedded into national and international laws as well as informal standards by respected nongovernmental organizations around the world. Legal and technical instruments, tools, and other measures to prevent occupational accidents and diseases have been put in place in all countries, albeit at different levels of comprehensiveness, sophistication, implementation, and enforcement capacity (Al-Tuwaijri et al., 2008).

The profound link between health at work and survival of human society is ubiquitous, timeless, and knows no geographic bounds. Civil society need not write new national and international occupational health laws in order to achieve societal goals of protecting workers, work, and their employers. As required under many statutory schemes, in-house occupational safety and health (OSH) compliance programs serve two major functions: (a) providing clear evidence of an enterprise's commitment to obeying the law, while (b) identifying hazards in time to reduce costs and potential liability (Feitshans, 2013).

Internationally, the law provides more than a yardstick for measuring compliance; however, OSH laws underscore the fundamental character of such health protections. Sound occupational health programs are universally necessary when promoting health and sustaining civil society. Occupational health laws also reflect an international consensus that protecting health at work and preventing occupational death is fundamental to commerce. Scientific precautionary principles are therefore embedded in many health laws, and the philosophy that work-related illnesses are an avoidable aspect of industrialization is a fundamental tenet of OSH laws. Despite the polemic outcry that OSH protections shackle innovation and industry, leading economists have ultimately accepted that some OSH regulation is unavoidable and necessary to prevent or compensate for market failures that cause unacceptably expensive harms (Viscusi, 2006). Therefore, the remarkable capacity building for occupational health infrastructure that has occurred in the late twentieth century, spanning into the first two decades of the twenty-first century, provides evidence that there are both practical and economic benefits derived from occupational health programming.

10.2   GLOBAL PERSPECTIVE: REGULATIONS AND STANDARDS SAVE THE LIFE OF EMPLOYERS

One major change in the cost-benefit economic equation of OSH programming is the cost of background research about techniques of risk assessment and risk mitigation itself. Open access to such resources, which transcends international boundaries, is available free of charge from a wide range of governmental agencies that did not exist a few decades ago (Hämäläinen, 2017). Ranging from well-respected websites such as the US National Institute of Occupational Safety and Health (2020) to electronic data sources of international and regional organizations, in particular the International Labor Organization (ILO), the World Health Organization (WHO), the European Union (EU), and the Association of Southeast Asian Nations, institutions, agencies, and public websites (Hämäläinen, 2017). Conversely, failure to create an in-house program is indefensible on grounds of economic costs in the event that the employer is challenged by an inspection resulting in noncompliance fines or penalties. These resources offer information that is practical and useful not only at home but also as the employer plans to procure supplies or send goods, services, merchandise, or completed products to partners and clients in other jurisdictions. A few hours spent reviewing the situation using information that is accessible free of charge in advance saves waste and prevents liability!

Assembling the relevant information, regardless of whether produced in-house by customized research or collected from sources available for free, is the first step in a process of designing an in-house compliance infrastructure that ultimately is a key tool for implementing the relevant laws. Compliance programs offer concrete proof of due diligence supporting a defense in case of inspection leading to citations for violations of law and in cases of litigation alleging liability. For new technologies and novel materials or unforeseen workplace risks from new hazards such as the pandemic of Covid-19, a sound in-house compliance program is the backbone of proving due diligence that can allow an employer to enjoy a presumption of having done the right thing in areas of the law where the risks are unquantified and the limits of otherwise reasonable actions to prevent harm are unknown (Feitshans, 1997). In-house occupational health and safety compliance systems can:

1. Avoid potential tort liability;
2. Reduce lost productivity and downtime;
3. Avoid OSH fines and penalties;
4. Prevent injury, illness, and disease;
5. Coordinate compliance with related laws such as environmental protection statutes thus enjoying double duty cost-savings when faced with inspections or allegations of liability;
6. Provide support for compliance with disability laws;
7. Enhance cooperation with governments for emergency preparedness and security;
8. Reduce worker compensation costs;
9. Reduce employer-based health insurance costs through targeted health promotion; and
10. Evince due diligence for privacy and security of data.

One measure of effectiveness for these procedures involves reduced quantifiable costs, such as decreased fines, penalties, insurance costs, downtime, and liability claims. Workers' compensation also creates a pervasive need to demonstrate to third parties in court or administrative hearings the due diligence of an employer or other responsible party regarding compliance programs that provide evidence of adherence to the standards under the law. The size and frequency of such claims vary greatly across different nations, depending not only on their economic level and socioeconomic conditions impacting the workforce but also on the availability of national health insurance and social determinants of health that support prevention before occupational health problems bloom into recordable injuries or illness. In this regard, free access to legal databases about occupational health in supplier nations or end-user locations is invaluable to any employer, large or small (ILO, 2020). OSHwiki has been developed by European Union-Occupational Safety and Health Administration (EU-OSHA) to enable the sharing of OSH knowledge, information, and best practices, in order to support the government, industry, and employee organizations in ensuring safety and health at the workplace. OSHwiki aims to be an authoritative source of information that is easily updated, edited, or translated and reaches beyond the OSH community (OSHwiki, 2020).

A difficult point for internal programming is the willingness to give candid scrutiny to embarrassing or expensive in-house problems, and the ability to gather sensitive information in-house in order to troubleshoot problems. Compliance staff must have the power and professional integrity to use sound occupational health information, even in the face of potential economic loss, from stopping or modifying a proposed course of action and the ability to use that power with accountability but without retaliation.

10.3 INTERNATIONAL LABOR ORGANIZATION GUIDELINES FOR CREATING, DESIGNING, AND IMPLEMENTING WORKPLACE SAFETY AND OCCUPATIONAL HEALTH PROGRAMMING

"The need for a global approach to OSH management was recognized as a logical and necessary response to increasing economic globalisation, while the benefits of systematic models of managing OSH became apparent as a result of the impact of ISO standards for quality and the environment" (Al-Tuwaijri et al., 2008). In 2017, there were an estimated 2.78 million fatalities compared to 2.33 million estimated in 2011 (Hämäläinen, 2017). For fatal occupational accidents, there were 380,500 deaths, an increase of 8% in 2014 compared to 2010. In 2015, there were 2.4 million deaths due to fatal work-related diseases, an increase of 0.4 million compared to 2011. In total, it is estimated that more than 7,500 people die every day; 1,000 from occupational accidents and 6,500 from work-related diseases. Refined methodology inevitably reveals more death and illness but also creates a more robust system for recording and analyzing reliable information. Paradoxically, the authors attributed increased deaths and fatalities in those 3 years to a higher underreporting estimate compared to the previous estimates, but this early effort at quantifying the magnitude of death and injury at work underscores its global significance.

10.3.1 THE ROLE OF THE ILO IN PROMOTING OCCUPATIONAL SAFETY AND GLOBAL HEALTH

In Geneva, Switzerland at the end of World War I, people understood the fundamental link between healthy working conditions, world peace, and the survival of all human society. As stated in the Preamble to the International Labor Office, the Constitution of 1919 states, "conditions of labour exist involving... injustice, hardship and privation to large numbers of people." The Preamble also notes, "universal and lasting peace can be established only if it is based upon social justice." And, significantly for the international mission of occupational health laws and their implementing programs, "the failure of any nation to adopt humane conditions of labour is an obstacle in the way of other nations which desire to improve the conditions in their own countries." (ILO, 1944). The ILO main headquarters is shown in Figure 10.1.

The ILO standards provide the basis for dynamic but feasible workplace health protection under the law. Since the founding of the ILO Committee of Experts on the Application of Standards in 1926, there has been international accountability for governments that fail to meet their obligations to maintain the standards they accepted by ratifying a convention. Governments who have violated key standards

FIGURE 10.1 International Labor Organization Headquarters in Geneva, Switzerland. (Photo courtesy of Thomas P. Fuller.)

are invited by the ILO to explain their problems at the annual International Labor Conference, held in June in Geneva, Switzerland every year since the recovery period after World War II. In the contemporary context, these values translate into the C187 Promotional Framework for OSH, which states,

1. Each member shall formulate, implement, monitor, evaluate, and periodically review a national program on OSH in consultation with the most representative organizations of employers and workers.
2. The national program shall:
   a. promote the development of national preventative safety and health culture;
   b. contribute to the protection of workers by eliminating or minimizing, so far as it is reasonably practicable, work-related hazards and risks, in accordance with national law and practice, in order to prevent occupational injuries, diseases, and deaths and promote safety and health in the workplace;
   c. be formulated and reviewed based on analysis of the national situation regarding OSH, including analysis of the national system for OSH;
   d. include objectives, targets, and indicators of progress; and
   e. be supported, wherever possible, by other complementary national programs and plans, which will assist in achieving progressively a safe and healthy working environment (ILO, 2006).

Drawing from the principles defined in the ILO Guidelines on OSH management systems, C 187 applies a similar approach to the management of national OSH systems reaffirming Convention No.155, C155- OSH Convention, 1981 (No. 155) (ILO, 1981) by providing an infrastructure for managing OSH, and establishing a prevention culture and progressively enhancing occupational health services. Article 4.1 states C155's goal of fostering the development of a "coherent national policy" concerning OSH protections. To this end, C155 obligates ratifying member states to promote research, statistical monitoring of hazardous exposures (such as medical surveillance measures, not unlike technical standards in the member states), and worker education and training. Consultation with representative organizations and employers is required before exemptions will be granted, and any exclusion for categories of workers requires reporting on efforts to achieve "any progress towards wider application" pursuant to Article 2.3. This flexible regulatory means that C155 can also adapt to the trend toward "personalized medicine" while encouraging ongoing national OSH data collection, risk management, and capacity building. Taken together, C155 and C187 operationalize precautionary principles using fundamental tenets of industrial hygiene.

Much of C155 reads like a checklist for a sound occupational health compliance program, except that the requirements apply to governments instead of employers. Governments that ratify this convention promise to include these components of risk management systems into their laws, or else they are not in compliance with the international labor standards. ILO Convention on Occupational Health Services (1985),

No. 161, and ILO Convention Number 170 1990 "Convention concerning Safety in the Use of Chemicals at Work" complement the provisions of Convention No. 155 that requires member states to establish competent governmental institutions that regulate and inspect workplaces (ILO, 1990). Under C155 member states commit to duties that involve surveillance of the work environment to identify, assess, prevent, and control occupational health hazards at the source. Under C155, "National competent authorities" designate authorized administrative agencies to ensure that they will create and implement an inspection system, to enforce national laws and regulations on OSH. C155, Article 3(e) offers the definition of health, "in relation to work, indicates not merely the absence of disease or infirmity; it also includes the physical and mental elements affecting health which are directly related to safety and hygiene at work." Thus, the ILO has adopted WHO Constitutional language to workplace settings, by making a strong assumption about the relationship between health and work. The text does not question the role of work in disease causation but simply states that the relationship exists.

Some of the features that are required for the design of a complete occupational health national regulatory scheme include inspection, development of national safety and health standards, and protection of worker rights to information, training, and complaints. These principles for information dissemination in the convention represent an international codification of the so-called "Right to Know" that Dr. Alice Hamilton and her peers at ILO worked hard to develop in the 1930s. National inspection services may also complement their activities with an advisory role on voluntary initiatives. Thus, a nationwide system should be in place in each nation consisting of a national system of recording and notification of occupational accidents and diseases, regularly updated for preventive purposes. Collecting statistical data about the incidence and prevalence of injuries and ill health due to accidents and exposures allows administrators to develop a clear picture of priorities for intervention. Occupational injury and disease compensation and rehabilitation systems can also use this information to offer experience/rating incentives for reduced injury. Oversight of such activities by governments is included in the role of the Committee of Experts on the Application of Standards, which meets at the International Labor Conference in Geneva, Switzerland every June.

C155 is an international template for the so-called "Right to Know" that is granted to workers and communities in civil society. This bundle of rights includes the right to be informed about the hazards, safe handling and use of dangerous materials, and access to working safety equipment free of charge; the right to be involved in the management and supervision of OSH measures at the workplace and the right to be organized in a representative group that can select delegates to OSH committees; the right to regularly scheduled updates concerning information and training on hazards/risks associated to their work and the measures to prevent them; the right to complain about impunity about unsafe circumstances; and the right to refuse hazardous work and not be required to return, in case of imminent serious danger to their health and life, without retaliation, with representation. In parallel, responsibilities of workers regarding such information require that workers follow safety and health rules when using protective equipment; participate in safety and health training and awareness-raising activities; cooperate with their employers to implement safety and health measures; and inform their direct supervisor if they withdraw from imminent and serious danger, stating their reasons.

10.3.2 PROGRAMS BY INDIVIDUAL GOVERNMENTS: COUNTRY PROFILES AT ILO BASED ON C155 CAPACITY BUILDING OF COMPETENT GOVERNMENT AUTHORITIES

The ILO offers many forms of technical partnership for member states. These efforts include assistance for drafting model legislation, onsite missions to solve national problems, and financial support for special projects. Stakeholders and the general public can access codified laws, ratified ILO Conventions, and a variety of technical information by using NORMLEX, the ILO database that organizes information into country profiles. The large number of nations participating in the self-reporting process for the database support the concept that occupational health rules and legislation exist throughout the world and that there is a global consensus about the importance of having occupational health programming under law.

10.3.3 COLLABORATION BETWEEN THE ILO AND THE WHO

The ILO/WHO Committee on Occupational Health, created to advance the purposes of C155 and the chemical safety convention, adopted a comprehensive definition of the aim of occupational health programming: "Occupational health and safety should aim at the promotion and maintenance of the highest degree of physical, mental and social well-being of workers in all occupations" (WHO, 2007). Part of this effort includes the WHO Declaration, Occupational Health for All that addresses occupational health for the very first time among WHO international instruments; the Declaration, although vague, makes a strong reference to the existing underlying human rights laws, such as the WHO Constitution, and offers the first insight to the remarkably broad scope of occupational health and the work of occupational physicians. Adopted in 1996, the Declaration amplifies the terms of the World Health Organization Health For All (WHO HFA2000) Plan for Action by answering the need to develop occupational health programs at a time when rapid changes in working life impacted the health of workers and the health of the environment in all countries of the world. Attended by 27 countries, plus WHO, ILO, United Nations Development Program (UNDP), and the International Commission on Occupational Health (ICOH), the Declaration adopted a proposal for action and implementation of its target goals. In particular, point 9 of the Declaration reaffirms each worker's "right to know the potential hazards in their risks in their work and workplace, including the development and use of appropriate mechanisms … in planning and decision-making concerning occupational health."

Justification for this collaboration is found in the International Covenant on Economic, Social, and Cultural Rights (ICESR), "Promotion of Industrial Hygiene" Article 12. Of all the UN-based international human rights documents, ICESCR Article 12 most clearly and deliberately addresses health. It has the clearest language of all international human rights instruments regarding "industrial hygiene" and protections against "occupational disease." (ICESCR, 1976).

10.4   HOW NATIONAL REGULATIONS ON OCCUPATIONAL SAFETY AND HEALTH ACTUALLY PROTECT WORKERS

In a complex mix of various states of economic development, different types of workplaces, and great differences in measuring health and safety outcomes, how can we know whether laws on the topic actually help a society?

10.4.1   MARKET IMPERFECTION ALONE DOES NOT ENSURE SUCCESSFUL INTERVENTION USING REGULATION

Leading proponents of deregulation or minimum regulation in the safety health and environmental context analyze the cost of avoiding or reducing risk. Under this view, if the cost of regulation to reduce risk is low enough, the regulation is justified. Economists agree on observing that the economic imperative for having consistent labels and requirements for material safety during the handling of dangerous materials provides an example where the need for information that workers and consumers cannot provide for themselves can outweigh ordinary political concerns. Sticking points in this analysis are the value assigned to human life from the standpoint of productivity and value across age and demographic variables. Based on a working assumption that imperfections in the market give rise to risks that require regulation, but that "imperfection alone does not ensure successful intervention."

10.4.2   WILLINGNESS TO PAY FOR SAFETY

Having demonstrated that risk reduction costs are more compelling than the uni-dimensional construct of valuation for human life when determining the economic basis for regulations, Vicusi is credited with having successfully settled a dispute between the US Occupational Safety and Health Administration (OSHA) and the US Office of Management and Budget in order to justify the acceptance of the hundreds of standards that came onto the OSHA agenda during the so-called PEL project, which formed the underpinnings of implementing the OSHA Hazard Communication Standard (OSHA, 1994). Viscusi claims that if markets functioned perfectly there would be no need for government risk regulation beyond moral or ethical obligations to address fairness and other variables that are not directly concerned with economics (Viscusi, 2006). By contrast, jurisprudential research synthesizing the link between the constitutional language protecting the health and a positive health impact on the lives of women and children are discussed in the WHO Expert report. "Women and Children's Health: Evidence of Impact of Human Rights" (Bustreo and Hunt, 2013) tries to prove that even if moral and ethical considerations could be quantified and applied to measures such as the global disease burden, economic arguments support safety and health protections. Constitutional protections are the ultimate form of regulation, and therefore, these two documents represent views at polar extremes. Ultimately, these arguments reflect different approaches to the willingness to pay to reduce risk, an aspect of economic policy that remains controversial even in the era of Covid-19.

Viscusi asserted that "market failures that give rise to a rationale for intervention are well known, not all market failures imply that market risk levels are too great. Hazard warnings policies often can address informational failures" (Viscusi, 2006). As Viscusi noted, the market failure of information surrounding the use of asbestos and the subsequent successful claims of employer liability proves that a market failure for information made it impossible for workers and consumers to make a meaningful choice in economic terms. In the late twentieth century, occupational health problems stemming from failure to warn about risk almost led to multinational bankruptcy (Rosenberg, 1985). Liability and stiff penalties, therefore, were justified as a counterweight to prevent this market failure from recurring. Asbestos is now the subject of massive global and national regulations, with the net result that asbestos use has increased worldwide with comparatively fewer injuries. deaths, and liabilities in several countries. One can only wonder whether the notorious asbestos exposures would ever have happened had there been a Globally Harmonized System for the Classification and Labeling of Chemicals, ("GHS") (UNECE, 2007) requiring disclosure and training in the safe use of dangerous materials a century ago. GHS is an indirect product of the asbestos experience. A multinational treaty whose implementation was crafted in partnership with industries, GHS provides unified symbols for coding materials as flammable corrosive or poison in a manner that can be understood across cultures languages, or customs borders. Following an international mandate, a coordinating group comprising countries, stakeholder representatives, and international organizations was established to manage the work.

10.2.3 EXAMPLE OF MAJOR ECONOMIC LIABILITY FOR RISKS: THE ROAD NOT TAKEN FOR ASBESTOS REGULATION AND ITS RESOLUTION

Some market failures may be exacerbated by government policies, and in those instances, safety and health as a policy driver take center stage becoming key to keeping a business alive. Asbestos offers one such abject lesson about the need for regulations to reduce risks compared to economic costs in the absence of regulation. The Government's failure to correct the informational market failure caused by asbestos exposure resulted in thousands of fatalities and billions of dollars of liability at the end of the twentieth century around the world. Surprisingly, the public health problem stemmed not from the inherent danger of the substance itself – society uses plutonium and radium, as well as a host of toxic substances and lethal machines on a daily basis with relatively little complaint. In the flood of liability claims that followed, multinational corporations who were powerful economic leaders came to the brink of becoming marginal employers when liability discussed in courtrooms around the world resulted in punitive damage awards (Kakalik, 1984).

In the crisis surrounding asbestos occupational deaths and the liability for recovery that followed, the asbestos companies themselves suffered from the absence of regulations and therefore ultimately welcomed the new national and international

(ILO, 1986) regulations that came into place to demonstrate both the minimum standard for safe care and the maximum penalties that could be extracted for failure to meet that standard, with liability for those companies who complied with the law becoming a problem of the past. National and international laws protecting OSH have succeeded in resurrecting a broken market response: international regulation of the asbestos exposure and the subsequent thriving of the asbestos industry. Internationally and in most nations, the law is clear about requirements to disclose dangers, their possible consequences, and the methods for reducing risk and preventing avoidable harm in a manner that simply did not exist when asbestos litigation began, nearly a century ago, when the dangerous exposures occurred. Regulatory procedures, combining disclosure during training with monitoring to reduce exposure have altered the course of occupational epidemiology within the asbestos industry, so that notoriously dangerous workplace risks do not impede flourishing commerce.

Balancing this economic policy dilemma has been addressed in international standards regarding occupational exposure to asbestos in 1986, ILO Asbestos Convention (C162) (ILO, 1996). This international standard exemplifies the international scientific consensus that asbestos exposure in the workplace is dangerous, requiring a regulatory apparatus designed to minimize risk while enabling its use at work. The format of the Convention adopts a flexible approach to asbestos regulation. Instead of setting a specific exposure limit, the standard enables national and local jurisdictions to adjust their requirements under law in order to meet the special needs unique to their situation. Although this approach has been criticized as offering "easy" standards for compliance in some nations and more rigorous standards in others, C162 offers a positive incentive for compliance among employers from a wide gamut of economic situations, thus saving lives and reducing hazardous exposures worldwide. Consistent with C155, the ILO Convention on workplace safety and health, the requirements for disclosure about health and safety hazards and training for the proper use, handling, and transport of asbestos, as determined under the auspices of national law, embody clear protection against the failure to warn of potential harms and codify the "Right to Know."

According to the ILO Convention,"(a) the term asbestos means the fibrous form of mineral silicates belonging to rock-forming minerals of the serpentine group, i.e., chrysotile (white asbestos), and of the amphibole group, i.e., actinolite, amosite (brown asbestos, cummingtonite-grunerite), anthophyllite, crocidolite (blue asbestos), tremolite, or any mixture containing one or more of these" (ILO, 1996). The key feature of a flexible framework is that although many important details of the program are not expressly stated, the key elements of a good program are included within its purview. In so doing, the regulatory structure that was created in partnership between the corporate decision-makers and government saved the life of the industry.

Getting governments to transition away from voluntary disclosure, and moving towards regulations to demonstrate full compliance with new occupational health and safety laws is not easy. A partnership between employers and the regulators that govern them is difficult but very important for creating workable programs that incentivize everyone to fulfill their duties as workers, employers' investors, and civil society stakeholders (Howard, 2009). The ILO Convention 162 Article 3. 1. allows great leeway to regulators to create doable programs: "National laws or regulations shall prescribe the measures to be taken for the prevention and control of, and protection of workers against, health hazards due to occupational exposure to asbestos. 2. National laws and regulations drawn up in pursuance of paragraph 1 of this Article shall be periodically reviewed in the light of technical progress and advances in scientific knowledge." (ILO, 1996). Previous ILO standards ratified before C162 concerning asbestos exposure reflected political agreement to limit toxic exposures at a certain level specified in a given Convention. By contrast, ILO C162 has no specific exposure indices. "Medical surveillance" of asbestos workers under ILO C 162 refers to monitoring health status. Under C162, Part IV, Article 21 "Surveillance of the Working Environment and Workers' Health" applies to all activities involving exposure of workers to asbestos. The ILO C162 Article 21 provides that medical surveillance shall be composed of five components: (a) medical examinations; (b) monitoring without cost to workers; (c) information and "individual advice" to workers regarding results of medical examinations (the so-called "right to know"); (d) alternative jobs for workers for whom asbestos exposure is "medically inadvisable"; (ILO, 1996) and (e) a national infrastructure for notification of disease within an overarching, cohesive administrative scheme for inspection, engineering controls, and addressing long-term health impacts of exposure (ILO, 1996).

Moreover, C162 states that at no cost to the worker medical examinations should take place during working hours, without loss of pay. Member states are free to determine the frequency, (e.g., annual or biennial) place, (e.g., at the worksite or in a governmental health facility), and extent of such examinations. Many nations require preplacement screening and periodic examinations (although the length of time between examinations may vary). Nations are allowed to determine the content of medical examinations, which may include: occupational and smoking history, physical examination, pulmonary function test, and a chest radiograph.

The strongest systems are cyclical and follow the commonly accepted industrial hygiene principles of occupational health and safety of anticipation, recognition, evaluation, and control. The degree of acceptable risk, the methods of risk assessment, and the measures of effectiveness for the same or similar hazards shall be determined by the context in which they arise, and therefore may differ depending upon circumstances. Standardized tools for this process within the context of a clear in-house compliance infrastructure feature need not be limited to auditing and the use of risk management tools that support proof of due diligence.

10.5 CONCLUSION: "I HAVE A SMALL COMPANY WITH JUST A FEW EMPLOYEES, AND WE DON'T HAVE MANY ACCIDENTS."

The profound link between health at work and survival of human society is ubiquitous timeless and knows no geographic bounds even in an age of nanotechnology addressing new diseases and illness (Feitshans, 2018). Civil society need not write new international laws if we courageously recognize and apply existing laws. As technology progresses, underlying basic human needs remain constant: Work health and survival have been inextricably linked throughout the history of human civilizations.

Without work, society cannot survive, and no work can perpetuate society without health. No society has survived without producing things or without work. We enjoy the fruits of many past civilizations today, as we draw upon their architecture such as the Pyramids, the Parthenon, and the Great Wall. Philosophies and values embedded in ancient cultures can touch our daily lives, even today. Indeed, the remnants that survive from ancient cultures are found in architecture, statues, and pottery artifacts of the skilled crafts and creative labors of lost societies. None of these types of work – the great monuments, writings, or arts – could exist without a modicum of human health. Civilizations can be brought to a halt in times of plague and pestilence; even the most impressive of collective efforts can be stopped when injuries overtake any individual's ability to work. Society, therefore, needs both working people and healthy people, in order for the civilization to survive. However, these classifications are not dichotomous or mutually exclusive. The fluid categories of sickness and health and economic well-being of employers and greater society fluctuate.

In conclusion, there is no dearth of international laws to provide a legal basis for implementing sound industrial hygiene practices and occupational health protections worldwide; there is no need for a popular hue and cry that there "ought to be a law" to protect people while working. International Laws governing occupational health and safety provide an important benchmark for measuring national laws. Economists who decry over-regulation nonetheless accept that informational regulation is necessary in order to prevent liability from market failures when people do not have enough accurate information to make good choices. Furthermore, fines and penalties must be high and paired with adequate inspections for laws to effectively incentivize employers, to obey laws, and to allocate resources to comply with safety and health laws (Viscusi· 2006). The relationship between the government and entities that are subject to regulation (Sigler, 1988) is complex, but establishing occupational health and safety requirements under the law can save the life of marginal enterprises as well as protect public health.

REFERENCES

Al-Tuwaijri, S., Fedotov, I., Feitshans, I., Gifford, M., Gold, D., Machida, S., Nahmias, M., Niu, S., Sandi, G., XVIII World Congress on Safety and Health at Work, Seoul, Korea, *Introductory report – Beyond death and injuries: The ILO's role in promoting safe and healthy jobs.* Geneva: International Labour Organization, June 2008. ISBN: 978-92-2-121332-1 (print). ISBN: 978-92-2-121333-8 (web pdf). International Labour Office occupational safety / occupational health / safety management / role of ILO 13.04.2

Bustreo, F., Hunt, P., *Women's and Children's Health: Evidence of Impact of Human Rights.* Geneva Switzerland: WHO Press, May 2013.

Feitshans, I., Designing an Effective OSHA Compliance Program, West/Thomson/Reuters and online Westlaw.com, 2013.

Feitshans, I., *Discussion of Compliance Programs Bringing Health to Work,* Emalyn Press, 1997.

Feitshans, I., *Global Health Impacts of Nanotechnology Law,* Panstanford, Singapore, 2018.

Hämäläinen, P., Takala, J., Kiat, T., Global Estimates of Occupational Accidents and Work-related Illnesses, Workplace Safety and Health Institute, Ministry of Manpower Services Centre, Singapore, 2017. www.wsh-institute.sg. RPT-2-2017, ISBN: 9789811148446.

Howard, J., Murashov, V., National Nanotechnology Partnership to Protect Workers, *J Nanopart Res* July 2009, 11(7) pp. 1673–1683.

ILO, 2020. ilo.org/dyn/normlex/en/, NORMLEX country profiles of laws and international agreements; NATLEX: National Legislation on Labour and Social Rights; LEGOSH Global database on occupational safety and health legislation; EPLex; Employment protection legislation database; Compendium of court decisions. Additional information is available free of charge in parallel in every nation and from regional organizations such as the EU, OAS, ASEAN and Organization for African Unity (OAU).

ILO, ILO Asbestos Convention C 162, 1986. Ratified by 35 Nations as of June 2017.

ILO, ILO Constitution, 1919 "Declaration of Philadelphia, 1944".

ILO, C187- Promotional Framework for Occupational Safety and Health Convention, 2006 (No. 187).

ILO, C161-Occupational Health Services Convention, 1985 (No. 161) https://www.ilo.org/dyn/normlex/en/f?p=NORMLEXPUB:12100:0::NO::P12100_INSTRUMENT_ID:312306 accessed April 5, 2021.

ILO, Chemicals Convention C 170, 1990 Convention concerning Safety in the use of Chemicals at Work (Date of coming into force: 04:11:1993.) Convention:C170, International Labour conventions and Recommendations 1977–1995, Volume iii, International Labour Office, Geneva, Switzerland, 1996, p. 337.

ILO, C155 - Occupational Safety and Health Convention, 1981 (No. 155), 1981, ilo.org/dyn/normlex/en/f?p=NORMLEXPUB:12100:0::NO::P12100_ILO_CODE:C155.

ICESCR, International Covenant on Economic Social and Cultural Rights (ICESCR), ICESCR Article 12 reads: *"The States Parties to the present Covenant recognize the right of everyone to the enjoyment of the highest attainable standard of physical and mental health. 2. The steps to be taken by the States Parties to the present Covenant to achieve the full realization of this right shall include those necessary for:...(b) The improvement of all aspects of environmental and industrial hygiene; The prevention, treatment and control of epidemic, endemic, occupational and other diseases....;"* International Covenant on Economic, Social and Cultural Rights Adopted and opened for signature, ratification and accession by General Assembly resolution 2200A (XXI) of 16 December 1966 entry into force 3 January 1976, in accordance with Article 27.

Kakalik, J., Ebener, P., Feitiner, W., Haggstrom, G., and Shanley, M., *Variation in Asbestos Litigation Compensation and Expenses*, Rand, The Institute for Civil Justice, Rand Publication Series, Santa Monica California, 1984.

NIOSH, *Pocket Guide to Chemical Hazards, Manual of Analytical Methods*, 5th edition, 2020. https://www.cdc.gov/niosh/chemicals/default.html. Accessed October 15, 2020.

OSHA, Hazard Communication Standard 29 CFR 1910.1200, U.S. Occupational Safety and Health Administration, 1994. https://www.osha.gov/pls/oshaweb/owadisp.show_document?p_id=10099&p_table=STANDARDS. Accessed October 15, 2020.

OSHwiki. 2020. https://oshwiki.eu/wiki/Main_Page.

Rosen, G., *A History of Public Health, Monographs on Medical History*, New York, 1958, p. 17.

Rosenberg, D., Book Review: The Dusting Of America: A Story Of Asbestos -- Carnage, Cover-Up, And Litigation, *Harvard Law Review*, 99, 1693. See also *Outrageous Misconduct: The Asbestos Industry On Trial*, By Paul Brodeur, Pantheon Books, New York, 1985, and by the same author *Expendable Americans*, 1976.

Sigler, J., Murphy, J., *Interactive Corporate Compliance: An Alternative to Regulatory Compulsion*, Quorum Books, Greenwood Press, Westport, Connecticut, 1988.

UNECE, The official text of the GHS, which was adopted on June 27, 2007, is available on the web at: http://www.unece.org/trans/danger/publi/ghs/ghs_rev00/00files_e.html.

Viscusi, W., Regulation of Health, Safety, and Environmental Risks, Discussion Paper No. 544, 2/2006, Harvard Law School, Cambridge, MA, Harvard John M. Olin Discussion Paper Series, www.law.harvard.edu/programs/olin_center/.

WHO, Worker's Health Global Plan of Action – Sixtieth World Health Assembly, World Health Organization, Geneva, 2007. https://www.who.int/occupational_health/WHO_health_assembly_en_web.pdf?ua=1. Accessed October 15, 2020.

11 International Trade Agreements and Their Impact on Worker Health and Safety

Barbara J. Dawson
Dupont

Garrett D. Brown
Maquiladora Health &Safety Support Network

Andrew Cutz
Health and Safety Matters and Associates

Paul Leonard Gallina
Bishop's University

David S. Rodriguez Marin
Innovare and Mexican Industrial Hygiene Association

CONTENTS

11.1   INTRODUCTION

The world's economy is interconnected and trade by wealthier countries with poorer countries is important to the poorer countries' development. However, concurrent with this economic development is the risk of people working in less than optimal conditions. The Workplace Safety and Health Institute (Hämäläinen et al., 2017) estimated that there were 2.78 million work-related deaths globally in 2017, the last year for which data are available. Of these deaths, 380,500 were attributed to occupational accidents, and the remaining 2.4 million were attributed to occupational disease. Many of the deaths occurred in developing countries, especially those in Africa and Asia, where the safety and health regulations are not as strong and/or enforcement of those that do exist are lacking.

Many countries do not have robust worker health and safety regulations. Some that do lack the resources to enforce them. Trade agreements typically require that countries follow their own labor standards but economics and politics often limit the successful implementation of the programs that would enable compliance with those standards. Workers in developing countries often lack adequate job health and safety training. Many industrial workers in these countries have an agricultural background which had different hazards and a different pace of work than industrial workplaces. In addition, these workers do not have the education to recognize the industrial hazards. Another concern is that developing countries have a poorer health status and are without adequate primary care (Jeyaratnam, 1998). This combination of poor education, inexperience, the lack of robust enforceable health and safety regulations, and poorer overall health leads to more occupational injury and illness. Access to trained safety and health resources that could help improve working conditions in developing countries is also limited.

Most trade agreements allow free trade, which usually means there are no tariffs, quotas, or other barriers to international trade. Trade agreements that regulate international trade between two or more countries have the potential to improve the working conditions by requiring worker protections as a condition of the agreement. These protections could be strengthened if the participating countries were required to have labor standards that reinforced basic human rights. Without these in place, the countries that don't protect their workers have an unfair trade advantage.

11.2   BACKGROUND

To appreciate the potential impact of trade agreements on improvements in worker health and safety, an understanding of the organizational structures that led to their development is helpful. Generally, as worker labor rights improve, worker health and safety protections also progress.

11.2.1   WORLD TRADE ORGANIZATION

The World Trade Organization (WTO), founded in 1995, has its roots in the General Agreements on Tariffs and Trade (GATT) signed in 1947 by 23 countries or territories. The GATT was limited to trade in goods and followed the formation of the United

Nations (UN), which came into existence in October 1945 (United Nations, 1945). Both the GATT and UN came into being just after World War II and reflected the idea held by the participating countries that an agreement would contribute to peace and stability in the world. The UN Charter recognized the need for all countries to maintain full employment with the belief that this is an essential condition to the expansion of international trade. The WTO oversees three major international trade agreements: the GATT (the WTO umbrella treaty for trade in goods), the General Agreement on Trade in Services, and agreements on Trade-Related Intellectual Property Rights (TRIPS) and several other multilateral trade agreements (WTO, 2020).

Some member countries do not believe that labor standards are within the purview of the WTO, but there have been debates about whether standards could be an incentive to improve workplace conditions. Historically, the WTO agreements have not dealt with labor standards. During the 1996 WTO Singapore conference, there was consensus that member countries were committed to recognizing core labor standards, but they should not be used for "protectionism," that is, to shield a country's domestic industries from foreign competition (WTO, 1996).

11.2 INTERNATIONAL LABOR ORGANIZATION

In existence since 1919, the International Labor Organization (ILO) was created by the Treaty of Versailles. In their centenary document, Rules of the Game (ILO, 2019c), the subchapter, *Building A Global Economy With Social Justice*, included the reflection, "What the ILO's founders recognized in 1919 was that the global economy needed clear rules in order to ensure that economic progress would go hand in hand with social justice, prosperity and peace for all."

In the preamble to its constitution (ILO, 1944), the ILO states the principle that workers must be protected from sickness, disease, and injury arising from their employment. The ILO adopted eight essential core standards during its International Labor Conference at its eighty-sixth session in Geneva on June 18, 1998, considered by the ILO to be the "fundamental" human rights. These are sometimes referred to as the "core labor standards."

The eight conventions are (ILO, 2019a):

1. Freedom of Association and Protection of the Right to Organize Convention, 1948 (No. 87)
2. Right to Organize and Collective Bargaining Convention, 1949 (No. 98)
3. Forced Labor Convention, 1930 (No. 29) (and its 2014 Protocol)
4. Abolition of Forced Labor Convention, 1957 (No. 105)
5. Minimum Age Convention, 1973 (No. 138)
6. Worst Forms of Child Labor Convention, 1999 (No. 182)
7. Equal Remuneration Convention, 1951 (No. 100)
8. Discrimination (Employment and Occupation) Convention, 1958 (No. 111)

As of January 1, 2019, there have been only 1,376 ratifications of these conventions with 121 still required to achieve universal ratification (ILO, 2019a). Out of 187 countries, 146 (78%) have ratified all the eight conventions; 13 (7%) have ratified seven of the conventions; 12 (6%) have ratified six of the conventions; 5 (4%) have ratified five of the conventions; (2%) have ratified four of the conventions; 1 (0.5%) has ratified three of the conventions; 2 (1%) have ratified two of the conventions, and 1 (0.5%) has ratified none of the conventions (ILO, 2019a).

In addition to the eight fundamental conventions, the ILO has adopted over 40 standards and over 40 recommendations and codes of practice that deal with occupational safety and health (OSH). These ILO documents and conventions provide a foundation for the protection of workers in trade agreements, although the ratification of the OHS conventions by countries is only a small percentage of the ILO membership (ILO, 2019a).

Unfortunately, none of the OHS conventions are included in the fundamental conventions described in the above paragraphs. In recent years, there has been an effort internationally to incorporate one or more of the OHS conventions into the core labor standards, so that trade agreements would reinforce the profile and importance of OHS issues (Alston, 2004).

11.3 TRADE AGREEMENTS WITH WORKER HEALTH AND SAFETY PROVISIONS

Globally, as of January 15, 2020, there are 303 regional trade agreements in place (see Figure 11.1). Many countries have flagged worker rights as an important issue during trade negotiations with some stakeholders believing they are important to protect developed country workers from perceived unfair competition and to raise labor standards in developing countries. There is some controversy about whether trade

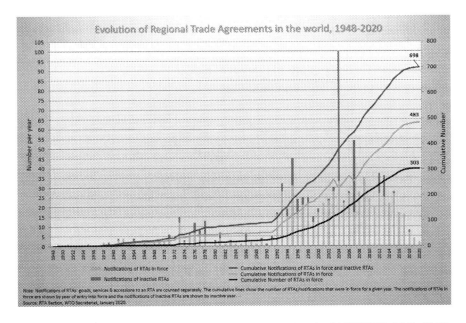

FIGURE 11.1 Regional Trade Agreements communicated to the GATT/WTO (1948–2020), including inactive RTAs, by year of entry into force. (WTO, 2020-with permission.)

agreements are the proper mechanism to accomplish this or whether that responsibility belongs to ILO or some other cooperative process. Both politics and economics impact trade agreements.

11.3.1 United States – Mexico – Canada Agreement

Sometimes referred to as North American Free Trade Agreement (NAFTA) 2.0, the United States – Mexico – Canada Agreement (USMCA) was signed by the United States, Mexico, and Canada in November 2018 with the revised version signed on December 10, 2019. It still required ratification by each of the countries. Mexico was the first to ratify the new agreement on June 19, 2019; the United States was next on December 19, 2019, and Canada ratified the agreement on March 13, 2020. The agreement also required each country to do additional work to prepare for the implementation of the agreement. On April 3, 2020, Canada informed the United States and Mexico that they had completed their internal processes and were ready to implement the agreement. Mexico made a similar announcement on April 9, 2020, and most recently, the United States notified the other countries that they were complete on April 24, 2020. Implementation of the agreement began on July 1, 2020.

The original NAFTA did not include labor provisions although a side agreement, called the North American Agreement on Labor Cooperation, was signed after-the-fact that included 11 guiding principles:

1. freedom of association and protection of the right to organize
2. the right to bargain collectively
3. the right to strike
4. prohibition of forced labor
5. labor protections for children and young persons
6. minimum employment standards, such as minimum wages and overtime pay, covering wage earners, including those not covered by collective agreements
7. elimination of employment discrimination based on race, religion, age, sex, or other grounds as determined by each country's domestic laws
8. equal pay for men and women
9. prevention of occupational injuries and illnesses
10. compensation in cases of occupational injuries and illnesses
11. protection of migrant workers

However, of these 11 principles, only 3 were subject to trade sanctions: labor protections for children, minimum employment standards, and prevention of occupational injuries and illnesses (Cimino-Isaacs and Villarreal, 2019). The agreement was intended to compel all three countries to ensure compliance with their respective labor regulations. Unfortunately, many US jobs were lost and went to Mexico, where there was cheaper labor. There was also a concern expressed by US labor unions that some manufacturing job wages in the United States were suppressed by employers threatening to move the jobs to Mexico.

Another criticism of the NAFTA side agreement – both labor and environmental – was that the process to formally complain about the failure of a signatory government to enforce its existing regulations was extremely lengthy, cumbersome, and not readily accessible for workers or communities (Public Citizen, 2019).

The hope is that the new USMCA will address the shortcomings of the original NAFTA and will strengthen labor provisions. One perceived improvement is that it not only requires the member countries to ensure compliance with their respective labor regulations but also requires the countries to adopt and maintain laws that support the ILO fundamental worker rights (Villarreal and Cimino-Isaacs, 2020). The agreement includes a Labor chapter meaning that labor obligations are part of the core of the agreement and therefore enforceable. However, in order to be considered a violation, trade or investment between the countries must be affected. The enforcement mechanism of the new agreement is through the initiation of two annexes (USA – Mexico) and (Canada – Mexico), which enable a "Rapid Response Labor Mechanism" that creates and authorizes an independent panel to conduct an inspection of any suspected violation. A panel of nine "experts in labor law and practice," three from both countries and the other three from outside of the two countries, will determine whether enough has been done to correct an issue. If the conclusion is that enough has not been done, sanctions may be applied (Lamonte et al., 2020).

The USMCA also includes new commitments relative to migrant worker protections, prohibitions against forced labor including child labor, workplace discrimination, and violence against workers. Other core labor standards addressed by the USMCA include freedom of association and the right to strike. Also incorporated into the agreement are minimum wage requirements that 40%–45% of the automotive content must be in jobs paying US$ 16/hour to avoid payment of a US tariff. This is the first attempt at including a guaranteed minimum wage in a trade agreement; however, missing are worker rights related to social justice, such as unemployment and job retraining benefits (Lamonte et al., 2019).

11.3.2 COMPREHENSIVE AND PROGRESSIVE AGREEMENT FOR TRANS-PACIFIC PARTNERSHIP

With its roots in the original Trans-Pacific Partnership (TPP) from which the United States withdrew in 2017, the Comprehensive and Progressive Agreement for Trans-Pacific Partnership (CPTPP) is an agreement between the remaining 11 countries including Australia, Brunei, Canada, Chile, Japan, Malaysia, Mexico, New Zealand, Peru, Singapore, and Vietnam. Prior to the withdrawal of the United States, the TPP was the largest free trade agreement. The agreement was signed on March 8, 2018, and went into force on December 30, 2018 (Amadeo, 2020), although not yet ratified by all of the signatory countries.

The CPTPP contains broad-ranging commitments on labor that reinforce internationally recognized doctrines that include:

- ensuring that national policies and laws provide protection for the fundamental principles and rights at work;
- preventing parties from deviation from their domestic labor laws to invite investment or trade;

- ensuring that laws provide protection for acceptable work conditions with respect to minimum wages, hours of work, and occupational health and safety; and
- encouraging collaboration and voluntary initiatives associated with labor matters.

The agreement also subjects labor provisions to the dispute settlement mechanism of the agreement. Thus, a party can impose trade sanctions if any of the commitments under the labor chapter are violated.

11.3.3 Comprehensive Economic and Trade Agreement

The Comprehensive Economic and Trade Agreement (CETA) is an agreement between Canada and the European Union (EU). Most of the provisions of CETA went into force on September 21, 2017. Its goals are to boost trade and generate growth and jobs (Government of Canada, 2020).

CETA removes most of the tariffs between Canada and the EU. It also includes a chapter on trade and labor with a provision that "The Parties recognise it is inappropriate to encourage trade or investment by weakening or reducing the levels of protection afforded in their law and standards." (CETA, Chapter 21). This chapter also requires both parties to promote public awareness of their labor laws and standards, enforcement, and compliance procedures. It goes further to require that information about these must be made available and that steps be taken to further the knowledge about them by workers, their representatives, and employers (European Commission, 2018).

11.4 CHALLENGES TO WORKERS' RIGHTS

The challenges to protect the rights of workers continue. Not only is there disagreement about what the labor standards should be but also enforcing those standards, once agreed upon, is difficult. Some of the disagreements are rooted in the economic realities of the developing countries and stem from different standards of living.

One example of a labor standard that on the surface seems morally the right thing to do concerns child labor. Prohibiting child labor to not allow children to be exposed to hazardous conditions and to enable them to pursue their education is embraced by some wealthier countries. However, in some developing countries, children's earnings may be an important part of a family's income. A standard prohibition of child labor could put an excess burden on a developing country, and some leaders of those countries fear that an enforceable prohibition on child labor could be used to protect businesses in developed countries from competition from those in developing countries. Others, however, believe that a child labor standard is essential for the protection of human rights (Burtless, 2001).

Labor standards make visible the difference in approach by different countries. Some countries, including the United States, prefer to see binding labor standards with a formal dispute process established. The EU approach, while viewing the standards as nominally binding, relies on stakeholder engagement, transparency, and public participation to hold the countries accountable (Elliott, 2014).

Enforcement mechanisms vary between agreements. Often, if one country has a trade complaint against another, the complainant country may retaliate against the offending country by withholding a trade benefit of similar value. This presents a challenge when there is a labor violation since it is difficult to calculate the benefit that the complainant country was denied.

Another mechanism that has been used to address labor standard violations is consumer boycotts. This practice isn't always effective since most consumers are reluctant to understand the conditions under which the goods are made and unwilling to pay the higher price for goods produced in countries that follow good labor practices.

11.5 CONCLUSIONS

The ethical implications of workers' rights will drive the evolution of labor standards. When a country embraces the ILO core standards, the workers should be able to negotiate for their rights including work hours, pay, benefits, etc. However, every country is not starting at the same place. Economics plays a critical role. There is not a quick solution to raising the minimum standard of living. As has been seen in the United States, higher wages lead to greater opportunities for education, which in turn leads to increased skills and productivity. A look at the past century shows that improvement in workers' rights is associated with better education and therefore better working conditions.

The moral question for health and safety practitioners is how long should we allow the natural progression to take? There's a conundrum of whether preventing countries without good labor standards from selling goods under a trade agreement will drive the change faster or hurt the welfare of the residents of those countries in the short term.

It is likely that some combination of binding and nonbinding labor standards with a progression of enforcement mechanisms will be around for a long time. Education of consumers and workers, along with increased transparency of the agreements and the processes used to communicate and enforce them, will drive the improvements in labor standards and therefore in worker health and safety protection.

REFERENCES

Alston, P. 2004. 'Core Labour Standards' and the Transformation of the International Labour Rights Regime. *EJIL*, 15: pp. 457–521.

Amadeo, K. 2020. Trans-Pacific Partnership Summary, Pros and Cons. https://www.thebalance.com/what-is-the-trans-pacific-partnership-3305581 (accessed July 13, 2020).

Burtless, G. 2001. Workers' Rights: Labor standards and global trade. https://www.brookings.edu/articles/workers-rights-labor-standards-and-global-trade/ (accessed July 13, 2020).

Cimino-Isaacs, C.D. and Villarreal, M.A. 2019. Worker Rights Provisions in Free Trade Agreements (FTAs). *Congressional Research Service*. https://fas.org/sgp/crs/misc/IF10046.pdf (accessed July 13, 2020).

Elliott, K. A. 2014. Worker Rights Provisions in Trade Agreements: Do They Matter? https://www.cgdev.org/blog/worker-rights-provisions-trade-agreements-do-they-matter (accessed July 13, 2020).

European Commission. 2018. CETA Chapter by Chapter, Chapter 23 – Trade and Labour. https://ec.europa.eu/trade/policy/in-focus/ceta/ceta-chapter-by-chapter/ (accessed July 13, 2020).

Government of Canada. 2020. Comprehensive and Progressive Agreement for Trans-Pacific Partnership. https://www.international.gc.ca/trade-commerce/trade-agreements-accords-commerciaux/agr-acc/cptpp-ptpgp/index.aspx?lang=eng (accessed July 13, 2020).

Hämäläinen, P., Takala, J., and Kiat, T., *Global Estimates of Occupational Accidents and Work-Related Illnesses*, Workplace Safety and Health Institute, Singapore (September 2017) ISBN: 9789811148446.

International Labour Organisation. 1944. Constitution. https://www.ilo.org/dyn/normlex/en/f?p=1000:62:0::NO:62:P62_LIST_ENTRIE_ID:2453907:NO (accessed July 13, 2020).

International Labour Organization. 2019a. Conventions and Recommendations. https://www.ilo.org/global/standards/introduction-to-international-labour-standards/conventions-and-recommendations/lang--en/index.htm.

International Labour Organization. 2019b. Ratifications by Country. https://www.ilo.org/dyn/normlex/en/f?p=1000:10011:8689583819630::::P10011_DISPLAY_BY:2.

International Labour Organization. 2019c. *Rules of the Game: An Introduction to the Standards-Related Work of the International Labour Organization*. Centenary edition. ISBN 978-92-2-132186-6. https://www.ilo.org/global/standards/information-resources-and-publications/publications/WCMS_672549/lang--en/index.htm (accessed July 13, 2020).

Jeyaratnam, J. 1998. Chapter 20, Development, Technology and Trade. *Encyclopaedia of Occupational Health and Safety*. ILO. 4th Edition. http://www.ilocis.org/documents/chpt20e.htm (accessed July 14, 2020).

Labonté, R. et al. 2019. USMCA (NAFTA 2.0): tightening the constraints on the right to regulate for public health. *Global Health*, 15: 44. https://www.ncbi.nlm.nih.gov/pmc/articles/PMC6518719/?report=reader (accessed July 13, 2020).

Labonté, R. et al. 2020. USMCA 2.0: a few improvements but far from a 'healthy' trade treaty. *Global Health*, 16: 43. https://www.ncbi.nlm.nih.gov/pmc/articles/PMC7201631/ (accessed July 13, 2020).

Public Citizen. 2019. 1994 NAFTA vs. Trump 2018 NAFTA 2.0 vs. 2019 New NAFTA: Labor and Environmental Standards and Enforcement. https://www.citizen.org/article/comparing-nafta-side-by-side/ (accessed July 13, 2020).

United Nations. 1945. Charter of the United Nations. https://www.un.org/en/charter-united-nations/index.html (accessed July 13, 2020).

Villarreal, M.A. and Cimino-Isaacs, C.D. 2020. USMCA: Labor Provisions. *Congressional Research Service*. https://crsreports.congress.gov/product/pdf/IF/IF11308 (accessed July 13, 2020).

World Trade Organization. 2020. Regional Trade Agreements. https://www.wto.org/english/tratop_e/region_e/region_e.htm (accessed July 13, 2020).

World Trade Organization. 1996. World Trade Organization: Ministerial Conference, Singapore, 9–13 December 1996. https://www.wto.org/english/news_e/pres96_e/wtodec.htm (accessed July 14, 2020).

12 Creating and Expanding the European Network Education and Training in Occupational Safety and Health (ENETOSH) to Build Capacity in OSH and Education

Ulrike Bollmann
Institute for Work and Health of the
German Social Accident Insurance

Dingani Moyo
Baines Occupational Health Services and
University of the Witwatersrand

Diana de Sousa Policarpo
Authority for Working Conditions (ACT)

Vinka Longin Peš
Croatian Health Insurance Fund

Ilda Luísa Figueiredo
Directorate-General for Education

Timothy David Tregenza
European Agency for Safety and Health (EU-OSHA)

Claus Dethleff
Headlog Multimedia

CONTENTS

12.1   INTRODUCTION

Advancements in technology and intensified globalization mean that networking is proving to be a new dimension for interacting and communicating. Relationships between people and organizations are increasingly based on horizontal interconnections (nodes) rather than on hierarchical systems with separate functional areas (Castells 1996). Networking encompasses an extremely broad spectrum, ranging from informal local initiatives to institutional cooperation through to global networks. This chapter focuses on networking in two areas in particular—the education sector and occupational safety and health (OSH). Experts from these areas traditionally find it difficult to communicate with one another on an equal footing and achieve results together.

The European Network Education and Training in Occupational Safety and Health (ENETOSH) was established to promote cooperation between OSH experts and education professionals. The main reasons for setting up ENETOSH are: (a) the importance of education in OSH policy has long been underestimated and not thoroughly understood and (b) safe and healthy working conditions are still a marginal issue for activities in the education sector today.

The following section describes the aim, approach, structure, and tasks of the ENETOSH network (Section 15.2), its four task forces (Section 15.3), and the strategy and milestones of ENETOSH (Section 15.4). Three examples of successful integration illustrate how capacities can be expanded through cooperation between

OSH and education at the national level (Portugal, Croatia) and supranational level (Africa) (Section 15.5). The final section (Section 15.6) takes a look at the network's future projects.

12.2 AIM, APPROACH, STRUCTURE, AND TASKS

The history of the network's establishment can be traced back to the EU Community Strategy on Health and Safety at Work 2002–2006. The underlying idea was to start integrating safety and health into education and training as early as possible. To do this, the European Agency for Safety and Health at Work (EU-OSHA) set up an expert group on "Mainstreaming OSH into Education"; held its first event under the Spanish EU Presidency in 2002; and, under the Italian EU Presidency, announced the Rome Declaration in 2003, which can be considered as a blueprint for integrating safety and health into education in Europe (EU-OSHA 2004, 135–138; Brück 2017). Between 2004 and 2013, numerous EU-OSHA reports were published on the subject (EU-OSHA 2004, 2007, 2009, 2010, 2011, 2013). In 2004, the Institute for Work and Health of the German Social Accident Insurance (DGUV) applied for funding to set up a network for OSH and education. The original establishment phase for ENETOSH from 2005 to 2007 was funded by the European Commission (Leonardo da Vinci Program 2005–146 253). Since 2005, the network has been coordinated by the IAG.

12.2.1 AIM

The aim of ENETOSH is to contribute to the integration of OSH into all levels of the education system. By doing this, safety and health should become a natural part of private, professional, and public life from early childhood and through all subsequent life phases. At the XVIII World Congress on Safety and Health at Work in Seoul in 2008, the term "culture of prevention" in the context of work was officially coined to describe this. Since the XXI World Congress in Singapore in 2017, the culture of prevention has been guided by the strategic goal of "Vision Zero".

12.2.2 APPROACH

ENETOSH follows a cross-policy approach by building a bridge between the policy areas of OSH and education. This is based on the assumption that OSH and education are interdependent. Safety and health is a basic prerequisite for good teaching and learning. Likewise, education and training make it possible for people to develop an awareness of the importance of their own safety and health and that of others and to have the necessary knowledge about safety and health when they first come into contact with the world of work. This enables future managers to acquire the competencies that will help them to shape tomorrow's world of work in a way that provides a decent work fit for human beings.

A unique feature of the ENETOSH network is its triple-track approach whereby ENETOSH addresses policymakers, researchers, and practitioners alike. This sets ENETOSH apart from other networks in which researchers mainly talk with other researchers, labor inspectors with labor inspectors, or where occupational hygienists promote the profession of occupational hygienists.

12.2.3 STRUCTURE

ENETOSH is a cooperation network, that is, a loose affiliation of complementary institutions with common interests which act as a whole or in subgroups and undertake initiatives via their connections (Tregenza 2014; Bollmann et al. 2018, 54).

ENETOSH started with 13 partners from 10 countries but has now grown to 97 members from 38 countries (84 institutions, 13 individuals). In addition to the numerous European Member States, ENETOSH has members from the USA, the Republic of Korea, India, Pakistan, the United Arab Emirates, Egypt, and Nigeria. Furthermore, the network has 25 ambassadors in 24 countries.

A memorandum of understanding has been signed with the following networks: European Network of Safety and Health Professional Organizations (2007); Robert W. Campbell Award, USA (2009); National Association of Organizations in Occupational Safety and Health of the Russian Federation (2012); International Occupational Hygiene Association (2018); and OSHAfrica (2019). In addition, ENETOSH has been working with the International Social Security Association Section on Education and Training for Prevention since 2006.

ENETOSH's work is overseen by a steering committee consisting of representatives from its members and social partners at European and national levels. The steering committee provides support to ENTOSH's coordinator and is responsible for the strategic framework of the work done by the network (see Section 15.4) (Figure 12.1).

12.2.4 TASKS

The work of the ENETOSH network follows the principle of content before structure—in other words, networking needs a purpose. The development of a competence standard for OSH trainers and instructors was the starting point for cooperation in the network (2005–2007). The competence standard was validated by a study conducted by the Dresden University of Technology (TU Dresden), Germany, and was further developed in the years 2008–2010.

One of the ongoing tasks for all members is to collect good practice examples (based on a set of quality criteria) of integrating safety and health into education. There are currently 1,002 good practice examples from 52 countries in the ENETOSH database. The majority are examples from continuing vocational education and training , followed by examples from day-care centers, schools, initial vocational education and training , and universities. Examples of education and training in occupational hygiene are mainly found in universities.

The work of ENETOSH was evaluated during its initial phase as an EU project (2005–2007). This evaluation was carried out by the Institute for Evaluation and Quality Development at Leuphana University, Germany. The process of establishing ENETOSH was scientifically monitored (internal evaluation) and the project results, as well as their perception from outside, were empirically assessed (external evaluation). Thus, right from the outset, there was clarity regarding the prerequisites for successful

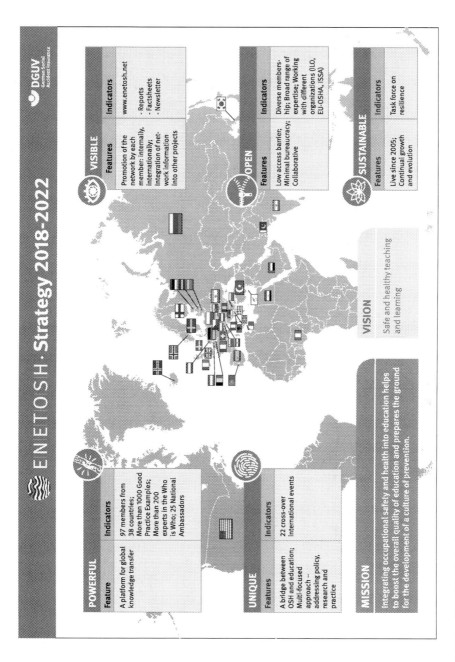

FIGURE 12.1 ENETOSH strategy.

networking (ENETOSH 2008). The following success factors identified at that time
have proven to be particularly important in the subsequent 15 years of networking:

- Good communication between network partners
- Mutual understanding and trust
- Low levels of hierarchy and bureaucracy
- Good balance between give and take
- Additional partners
- Development of the idea behind the network based on a long-term approach.

The database of good practice examples was systematically evaluated 10 years after
it was set up as part of an empirical study in 2018. The study was initiated by the
International Labour Organization (ILO) and funded by the British Safety Council.
Using the settings-based approach of the World Health Organization, networking,
participation, and sustainability were identified as key elements of successful integration of safety and health into education systems.

The following three trends for integrating safety and health into education systems were identified at the end of the study (Bollmann et al. 2018, 80):

1. The increasing importance of the UN's Sustainable Development Goals
   (SDGs) for mainstreaming OSH into education
2. The targeted implementation of good models in an international context
3. The systematic monitoring of the implementation of good models through
   translational research.

12.3 ENETOSH TASK FORCES

The work of ENETOSH is divided into subgroups known as task forces. This is
in line with the results of an ILO study on OSH networks (Rantanen 2018, 24).
This study identifies the model created by the Partnership of European Research in
Occupational Safety and Health as a good example for OSH networks. The following
provides a brief overview of the four task forces (Sections 15.3.1–15.3.4).

12.3.1 TASK FORCE: DISSEMINATION AND IMPLEMENTATION OF THE
GOOD HEALTHY SCHOOL MODEL INTERNATIONALLY

This task force is the "heart" of ENETOSH. ENETOSH pursues a holistic approach to
the integration of safety and health into the education system. This takes into account
a person's complete physical, mental, and social wellbeing, together with an ecologi-
cal perspective on OSH and education. The task force is led by the statutory accident
insurance for the public sector in North Rhine-Westphalia and Leuphana University,
Germany. The task force is responsible for bringing the Good Healthy School Model
to the network. The model was developed in Switzerland at the beginning of this mil-
lennium and was further developed in Germany (Braegger and Posse 2007; Paulus
2009; Hundeloh 2012; DGUV 2013). The main purpose is to improve the overall quality of education by systematically integrating safety and health into the organizational

development of schools. The main target group of the task force consists of school principals, teaching staff, and students. The focus is on the salutogenic management of schools (Paulus and Hundeloh 2020). The issue of mental health in educational institutions is an integral part of the Good Healthy School Model.

As part of a research project, the Good Healthy School Model is being disseminated internationally and its implementation in educational institutions is being scientifically monitored. The aim of assessing its implementation is to provide practical evidence of the success of the model.

12.3.2 TASK FORCE: MAINSTREAMING OSH INTO HIGHER EDUCATION

This task force is currently the "driver" for further developing ENETOSH. The work of this group focuses on the issue of how to integrate safety and health into higher education and how to empower the decision-makers of tomorrow to shape decent work now and in the future.

In cooperation with an international consortium led by Heinrich Heine University Düsseldorf, Germany, and in collaboration with the Karolinska Institute, Sweden, the task force has launched a project for Sustainable Development Goal 8 entitled "Promoting decent work and productive employment through higher education". The project has been approved by the Global Occupational Safety and Health Coalition, which is coordinated by the ILO.

After putting together a network of experts on the topic, the project will initially focus on finding curricula that integrate decent work and sustainable economic growth into university teaching. This will also look at which competencies are necessary for future managers. The next step is to address the issue of good working conditions in universities. This ENETOSH task force is led by the University of Girona, Spain, and the University of Brighton, UK.

12.3.3 TASK FORCE: RESILIENCE OF NETWORKING

This task force is the "*vade mecum*" of ENETOSH. It is an essential provider of advice and deals with the sustainability of the network.

Resilience is understood here as "the ability of individuals, communities, organizations or countries exposed to disasters, crises and underlying vulnerabilities to anticipate, prepare for, reduce the impact of, cope with and recover from the effects of shocks and stresses without compromising their long-term prospects" (IFRC 2014).

For ENETOSH, resilience is a question of how the network survives and maintains its relevance in the face of change.

Potential threats to ENETOSH were identified as being a loss of financial resources, loss of membership (particularly key network members), and a loss of relevance (either of ENETOSH in the field of education and training for OSH, or of the topic as a whole in the scope of education and employment policy).

These threats are interrelated. A declining membership means a loss of financial resources, loss of reach, and a loss of influence in the topic. Any network that is overly reliant on a central hub, whether a funding source or individual carrying out the tasks, is at risk should that central hub be removed.

ENETOSH has to achieve a virtuous circle whereby it ensures the topic of OSH education and training remains on the agenda of both employment and education policymakers, increasing the network's visibility and influence in the subject and by doing so attracts more membership among whom the costs and work can be distributed.

To do this, ENETOSH needs to maintain contact with and visibility to European and international policymakers and practitioners and be able to respond rapidly to their needs, using its unique resources and reach to provide current, relevant, useful, and accurate information.

12.3.4 Task Force: Elevate the Policy Profile of Mainstreaming OSH into Education

This task force represents a "touchstone" for the work of ENETOSH: Is it possible to sustainably embed the topic of mainstreaming OSH into education in national OSH and education policies? What conditions are necessary for successful cooperation between experts from different policy areas? What are the specific results of such cooperation?

The task force, currently, has two examples of how safety and health can be integrated into the structures and processes of national OSH and education systems (see Sections 15.5.1 and 15.5.2). The task force is led by the Authority for Working Conditions (ACT), Portugal, and the Croatian Health Insurance Fund.

12.4 STRATEGY AND MILESTONES

ENETOSH is an open and vibrant network. The work of ENETOSH is sustained by the commitment of its members and friends. The clearly defined aim of integrating OSH into education is a steady point of reference for the joint work done by the network.

The foundation for ENETOSH's continuous further development is a strategic framework that was developed from the findings of two-member surveys. ENETOSH's strengths, weaknesses, opportunities, and threats (SWOT) were systematically analyzed in 2014 and 2018. This also involved checking whether the network's goals are consistent with those of its members. The current strategic framework was approved by the ENETOSH Steering Committee in April 2018 and runs until 2022. The strategic framework also serves as a means of evaluating new trends and making it transparent to members how they are reflected in ENETOSH's objectives. This applies both to the inclusion of the concept of a culture of prevention in the ENETOSH objectives (2014) and the alignment with the strategic goal of Vision Zero (2017).

In 2019, ENETOSH adjusted its work to the context of the SDGs of the United Nations. The focus here is on linking SDG 8 "Decent work and economic growth" and SDG 3 "Good health and well-being" with SDG 4 "Quality education" in higher education. This is a way for the network to intensify its efforts with respect to this specific level of the education system (Figure 12.2).

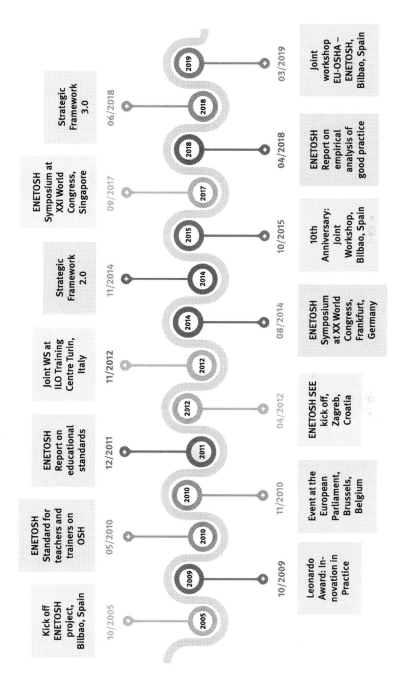

FIGURE 12.2 Milestones of the work of ENETOSH.

12.5   ENETOSH EXAMPLES OF SUCCESS

The following sections use examples from Portugal (15.5.1) and Croatia (15.5.2) to illustrate the framework in which different policy areas interact in such a way that OSH can be sustainably integrated into the education system. Chapter 15.5.3 looks at the collaboration between the OSHAfrica and ENETOSH networks.

12.5.1   PORTUGAL

The Ministry of Education of Portugal designed the document "Student Profile by the End of Compulsory Schooling" in 2017 (Order No. 6478/2017, of July 26) as a reference document for organizing the entire education system and for the work of schools. It supports convergence and coordination of decisions regarding the various dimensions of curriculum development. This profile is based on principles, values, and a clearly defined vision reflecting social consensus. It aims to ensure educational success for all students, reduce early school leaving, and promote the quality of inclusive education by enabling lifelong learning and encouraging curriculum development and citizenship education in an integrated manner.

The National Strategy for Citizenship Education (ENEC), launched in October 2017, was implemented in alignment with the "Student Profile by the End of Compulsory Schooling" as part of Decree-Law No. 55/2018 (Presidency of the Council of Ministers 2018).

The goals of the citizenship and curriculum development component are active citizenship; the development of competencies for a culture of democracy; and learning with regard to individual civic attitudes, interpersonal relationships, and social and intercultural relations. Citizenship and development is offered in the various education and training courses which cover 17 interconnected domains, including the world of work.

The domain of the world of work, together with the other domains, contributed to the operationalization of the "Student Profile by the End of Compulsory Schooling". From the perspective of lifelong learning, education for the world of work develops in diverse, interconnected, and complementary fields of education that include formal, nonformal, and informal education.

In accordance with the ENEC, and particularly in view of drawing up guidance documents to include the world of work, Guidelines for Education for the World of Work are being developed. The project is coordinated by the Directorate-General of Education (DGE) in partnership with the ACT, the ILO, Lisbon, the Institute of Employment and Vocational Training, IP, and the National Agency for Qualification and Vocational Education . The issue of safety and health at work is one of six global issues identified as being important to address with an evolutionary approach that covers the entire spectrum of education from preschool through secondary education.

In addition, these guidelines are in line with the National Strategy for Health and Safety at Work (Resolution of the Council of Ministers No. 77/2015, of September 18) (ENSST 2015–2020), which defines the policy of risk prevention and the promotion of well-being at work. Its first global objective is to develop and implement public policies for OSH. The first strategic objective is "Mainstreaming occupational safety and health into the education system - National program for the inclusion of occupational safety and health issues in school curricula at all levels of education". The second strategic objective is "Promoting the training of teachers, including other school staff, in occupational safety and health".

In-service teacher training courses and other initiatives to raise teachers' awareness of workplace issues (particularly safety and health) and to enable them to adopt participatory forms of working with students that encourage pluralistic and responsible participation by all are being planned and, in some cases, are already underway.

Through the ERASMUS+ Project "Mind Safety–Safety Matters!" (2015–2018 and 2018–2021), ACT ran five accredited teacher training courses across the country. This training, entitled "OSH Education and Safety Culture", involved 75 teachers responsible for 4,056 students across all levels of education, thus meeting the second objective of ENSST 2015–2020.

The second phase of this project will pursue and refine the objectives outlined by a consortium of seven partners from five countries: Youth in Science and Business Foundation, Estonia; University of Girona, Spain; Delft University of Technology, The Netherlands; and ÇASGEM, Turkey—coordinated by ACT and the Universities of Aveiro and Minho, Portugal. The project includes a "European Teaching Guide for OSH Training" for project-based teaching and learning, and two digital platforms under the umbrella name "OSH! What a bright Idea!" One platform targets teachers and provides them with classroom and teacher-training materials to prepare future workers for future occupational risks. The other platform targets young workers and contains an interactive booklet for students, including blind and partially sighted students and minorities.

Networking has never been more imperative for improving safety and health for all than it is today. In Portugal, several public and private bodies, social partners, and institutional partners have taken the first step toward achieving this goal—the dissemination of occupational safety and health concepts across all levels of education, starting now.

12.5.2 CROATIA

The Croatian Ministry of Science and Education developed the "Dual Education" experimental program as part of the reform of vocational education in accordance with the Republic of Croatia's Government Program for the period 2016–2020 (ASOO 2016). This part of the program, which refers to the education system harmonized with the needs of the labor market, is the basis for the legislative framework of the new Law on Dual Education which is in the process of being adopted.

Dual education is a form of vocational education and training that combines learning at a vocational education institution with learning in a company (see ASOO 2018). Participants in the dual educational process are students, vocational education institutions, and companies who expressed interest in participating in experimental implementation. The emphasis is on high-quality cooperation between the education sector and the world of work, with a focus on competency development, and in accordance with the principles of partnership, a guarantee of quality and standards, availability, flexibility, and the ability to adapt to change.

Companies supply the physical environment and resources needed for work-based learning, provide a mentor to the student, offer a stimulating and safe environment, carry out mandatory occupational safety measures during the performance of work-based learning, pay the student a contracted monthly wage, cooperate with the institution for vocational education, provide training for vocational school teachers, and enable the exchange of knowledge and new technologies.

The aims of dual education are to acquire knowledge, skills, and responsibilities for entering the world of work; develop an awareness of the importance of work in a safe manner; promote teamwork and partner cooperation; develop critical thinking while respecting other thoughts and ideas; and promote social justice and affirmative relationships to vulnerable groups.

The experimental implementation of dual education began in the 2018/2019 school year. An operational team was formed of representatives from the Ministry of Science and Education, the Ministry of Economy, Entrepreneurship, and Crafts, the Ministry of Labor and the Pension System, the Ministry of Finance, and the Ministry of Health. The Ministry of Science and Education team monitored the results of the experimental implementation and evaluated the implementation of the new vocational curriculum.

The experimental implementation was financed from the state budget and the project "Modernization of VET Programs" as part of the Swiss–Croatian cooperation program. There is also an investment for the regional centers of competence from the European Structural and Investment Funds to support vocational education reform.

The establishment of centers of competence in Croatia is in line with the Strategy for Education, Science, and Technology (Government of the Republic of Croatia 2014); the Vocational Education and Training System Development Program 2016–2020 (ASOO 2016); and the 2018 amendments to the Vocational Education and Training Act (DG EDUC 2018a; EACEA 2020). The centers of competence offer VET to students, as well as professional guidance and continuous professional development and training to professionals, teachers, and workplace mentors. Their main features include innovative learning opportunities, teaching and workplace mentoring excellence, state-of-the-art facilities, and intensive cooperation with local enterprises and other VET stakeholders.

The results of the Ministry of Science and Education's evaluation show that the implementation was successful and that more than 70% of the students, employers, and teachers were satisfied with the implementation of this experimental program. More than 90% of students achieved the planned learning outcomes at school and in the workplace with the employer.

By developing the dual education system according to its own specifics, in favor of successful cooperation between the economy and education, Croatia is preparing for a new period of challenging times, while remaining strongly aware of the need for digital transformation, research, and innovation for future jobs.

12.5.3 OSHAFRICA

OSHAfrica is a Pan-African OSH professional organization with over 540 members from 40 African countries. OSHAfrica was established with the purpose of bringing together all OSH professionals across Africa for collaborative work, sharing data, and improving workplaces across Africa. To date, it is one of the largest and most influential continental OSH bodies. Its board is gender-balanced, and all the subregions of Africa are well represented. As an influential body, OSHAfrica has forged strategic partnerships at both continental and international levels. Within Africa, OSHAfrica has formed partnerships with the African Union and the Pan African Virtual and E-University. At the international level, OSHAfrica works with key OSH stakeholders such as the ILO, ENETOSH, Workplace Health Without Borders, and many others.

OSHAfrica seeks to lead OSH research and studies across the African continent and to collaborate with external OSH organizations and agencies in developing OSH projects, programs, and data in Africa. As an inaugural continental body, OSHAfrica aims to bring together OSH professionals and associations across Africa and create an enabling environment for collaborative work and data sharing for professional improvement.

Human resources capacity development remains a major challenge for Africa. There are few institutions in Africa involved in the training of OSH professionals and practitioners. The lack of expertise in this area has impeded growth in OSH in most African countries. OSHAfrica has established three scientific committees that have done significant work in these areas to drive improvements in OSH in Africa. Through its scientific committees, OSHAfrica aims to improve OSH practice through research, training, and legislative formulation. The Scientific Committee on Education and Competency Development is mandated with developing strategic ways of improving the competencies of OSH practitioners and professionals. To overcome the deficits in OSH research and publications in Africa, the Scientific Committee on Data, Research, and Publication is charged with collating and analyzing OSH data across Africa. One of the key factors in improving OSH is in legislation and policy frameworks. The Scientific Committee on Legislation and Policy Formulation is mandated with the review of national OSH policies and related OSH legal instruments. OSHAfrica assists African countries in developing their OSH frameworks at national and regional levels.

It is against this background that OSHAfrica has strategically partnered and collaborated with key institutions both continentally and abroad. Partnerships with ENETOSH, the African Union, and other influential institutions have played a pivotal role in the improvement of OSH in Africa.

12.6 PROMOTION OF OCCUPATIONAL HEALTH—A NEW AREA FOR NETWORKING OF NETWORKS

The corona pandemic is one of the biggest challenges for safety and health worldwide. This is especially true for educational institutions.

The word hygiene (derived from the name of the Greek goddess of health known as Hygeia, whose mission was to maintain good health and prevent disease [IOHA n.d.]) is increasingly the focus of attention for policymakers, researchers, and practitioners in all sectors of society. As a result of the coronavirus, safety and health now seem to be integrated as a matter of course into everyday life and working life—a culture of prevention has become a reality. The Organization for Economic Cooperation and Development (OECD's) announcement that the next Program for International Student Assessment (PISA) study will, for the first time, also examine the social and emotional skills of schoolchildren seems to be a positive sign. The next PISA school study will focus on the relationship between teachers, students, and schools and how this has changed as a result of the corona pandemic (Schleicher 2020; cf. the design of the study planned so far in OECD 2018).

The question of which competencies are necessary in the age of networked human relationships, and thus in the age of a culture of prevention, is now relevant to all areas of society (cf. Bollmann and Boustras 2020, 4).

Corona is also a test case for mainstreaming OSH into education since the following needs are no longer perceived as just optional but rather, broadly speaking, essential:

* Innovative approaches to education (e.g., interactive e-cases, e-learning, curated blended learning; participatory webinars and workshops, AI-based corporate training)
* New outcome measures (e.g., competence as perceived self-efficacy—the belief that you can make a difference through your own actions (Bandura 1997; van Dijk 2017; Bollmann et al. 2018, 26, 35))
* The professionalization of education experts and occupational hygienists, occupational physicians and safety professionals with regard to good communication, salutogenic leadership, true participation, new forms of virtual and face-to-face teaching and learning, and interpersonal and institutional trust as the foundation of a culture of prevention (Bollmann and Boustras 2020, 5; Bollmann et al. 2020)

The challenge posed by a health crisis such as the corona pandemic also highlights the need for stronger links with health policy. The result of the evaluation of the ENETOSH good practice database in 2018 showed that 57% of the examples were assigned to the field of safety (Bollmann et al. 2018, 25). This result signaled a need for action, since health has traditionally been much better represented in education than safety.

In connection with EU-OSHA's Healthy Workplaces Campaign 2020–22 to tackle work-related musculoskeletal disorders , ENETOSH has been more involved in the fields of occupational health and public health since the second half of 2018. The focus is on promoting good musculoskeletal health in children and young workers. The approach is multidisciplinary and is aimed equally at policymakers, researchers, and practitioners in the fields of OSH, education, and public health. A new feature of the joint launch event with EU-OSHA in May 2019 was the body image concept (Lewis-Smith and Sherman 2019), which assumes that young people's self-perception of their body has a strong influence on their health. The follow-up event, planned for March 2020, was to focus on the importance of physical activity for education and health and aimed to strengthen interactions between networks and organizations in the health and education sectors. Examples include EULAR Young PARE, European Association for the Study of Obesity , Schools for Health in Europe, European Parents Association, Association of Teacher Education in Europe,

European Federation of Education Employers , European Trade Union for Education, European Student Unions . The event had to be canceled due to the corona crisis and is now planned as a webinar.

ENETOSH, in cooperation with an international consortium, has set itself the ambitious goal of showing where and how decent work and health and wellbeing are linked with educational quality in the field of tertiary education with the project "SDG 8- Promoting decent work and productive employment through higher education". Through this project, ENETOSH is addressing the area of education in which it has been the most difficult to integrate OSH. However, the framework of the United Nations' SDGs and embedding the project as a task group within the Global Occupational Safety and Health Coalition are showing promise.

REFERENCES

ACT - Authority for Working Conditions. 2016. *National Strategy for Health and Safety at Work (ENSST) 2015–2020. For a safe, healthy and productive work.* Available at http://www.act.gov.pt/(pt-PT)/SobreACT/DocumentosOrientadores/PlanoActividades/Documents/National%20Strategy%20for%20Health%20and%20safety%20at%20work%202015-2020.pdf (accessed May 20, 2020).

ACT - Authority for Working Conditions. 2019. *OSH! What a bright idea!* http://osh.act.gov.pt/ (accessed May 20, 2020).

ASOO - Agency for Vocational Education and Training and Adult Education. 2016. *Vocational Education and Training System Development Program 2016–2020.* Available at https://www.asoo.hr/UserDocsImages/VET_Programme_EN.pdf (accessed May 19, 2020).

ASOO - Agency for Vocational Education and Training and Adult Education. 2018. *Experimental Program Dual Education.* Available at http://www.refernet.hr/en/news/croatian-news/experimental-program-dual-education/ (accessed May 19, 2020).

Bandura, A. 1997. *Self-efficacy: The exercise of control.* Basingstoke: Worth Publishers.

Bollmann, U., R. Gründler, and M. Holder. 2018. *The integration of safety and health into education. An empirical study of good-practice examples on www.enetosh.net.* Berlin: German Social Accident Insurance.

Bollmann, U. and G. Boustras. 2020. Introduction. In U. Bollmann and G. Boustras (Eds.), *Safety and health competence. A guide for cultures of prevention*, 1–21. Boca Raton: CRC Press, Taylor & Francis Group.

Bollmann, U., Y. Lee, Y. Seo, H. Paridon, T. Kohstall, A.-M. Hessenmöller and C. Bochmann. 2020. The development of a model and a set of leading indicators for promoting occupational safety and health. *Prevention Science*, Special issue on "Culture of Prevention" (manuscript, accepted).

Braegger, G. and N. Posse. 2007. *Instrumente für die Qualitätsentwicklung und Evaluation in Schulen.* Bern, Switzerland: hep Verlag.

Brück, C. 2017. Mainstreaming OSH into education. Available at OSHWIKI https://oshwiki.eu/wiki/Mainstreaming_OSH_into_education (accessed May 19, 2020).

Castells, M. 1996. The rise of the networks society. *The information age: Economy, society, and culture.* Vol I. Oxford: Wiley-Blackwell.

DG EDUC - Directorate-General for Education, Youth, Sport and Culture. 2018a. *Education and Training Monitor 2018: Croatia.* Luxembourg: European Commission. Available at https://www.azoo.hr/userfiles/dokumenti/et-monitor-report-2018-croatia_en.pdf (accessed May 19, 2020).

DG EDUC - Directorate-General for Education, Youth, Sport and Culture. 2018b. *Education and Training Monitor 2018: Portugal.* Luxembourg: European Commission, ISBN

978-92-79-89867-9. Available at https://ec.europa.eu/education/sites/default/files/document-library-docs/et-monitor-report-2018-luxembourg_en.pdf (accessed April 6, 2021).

DGUV – German Statutory Accident Insurance. 2013. *Strategic concept: Using health to develop good schools*, ed. Specialist Division Educational Institutions. Berlin: German Social Accident Insurance.

EACEA - Education, Audiovisual and Culture Executive Agency. 2020. *Croatia: National Reforms in Vocational Education and Training and Adult Learning.* Available at https://eacea.ec.europa.eu/national-policies/eurydice/content/national-reforms-vocational-education-and-training-and-adult-learning-11_en (accessed May 19, 2020).

EACEA - Education, Audiovisual and Culture Executive Agency. 2020a. *Portugal: National Reforms in School Education.* Available at https://eacea.ec.europa.eu/national-policies/eurydice/content/national-reforms-school-education-53_en (accessed May 19, 2020).

ENETOSH – European Network Education and Training in Occupational Safety and Health. 2008. Evaluation of the network, 3, In U. Bollmann (Ed.), *ENETOSH Newsletter No 6.* Dresden: German Social Accident Insurance.

EU-OSHA - European Agency for Safety and Health at Work. 2004. *Mainstreaming occupational safety and health into education.* Report. U. Bollmann (Ed.), Available in six languages. Luxembourg: Publications Office of the European Union.

EU-OSHA - European Agency for Safety and Health at Work. 2007. *OSH in figures: Young workers – Facts and figures.* Report. E. Schneider and S. Copsey (Eds.). Luxembourg: Publications Office of the European Union.

EU-OSHA - European Agency for Safety and Health at Work. 2009. *OSH in the school curriculum: requirements and activities in the EU Member States.* Report. K. Sas and S. Copsey (Eds.). Luxembourg: Publications Office of the European Union.

EU-OSHA - European Agency for Safety and Health at Work. 2010. *Mainstreaming occupational safety and health into university education.* Report. S. Copsey (Ed.). Luxembourg: Publications Office of the European Union.

EU-OSHA - European Agency for Safety and Health at Work. 2011. *Training teachers to deliver risk education – Examples of mainstreaming OSH into teacher training programmes.* Working paper. S. Copsey (Ed.). Luxembourg: Publications Office of the European Union.

EU-OSHA - European Agency for Safety and Health at Work. 2013. *Occupational safety and health and education: a whole school approach.* Report. S. Copsey (Eds.). Luxembourg: Publications Office of the European Union.

Government of the Republic of Croatia. 2014. *Strategy of Education, Science and Technology.* Available at https://vlada.gov.hr/highlights-15141/archives/strategy-of-education-science-and-technology-nove-boje-znanja/17784 (accessed May 20, 2020).

Hundeloh, H. 2012. *Gesundheitsmanagement an Schulen. Prävention und Gesundheitsförderung als Aufgaben der Schulleitung.* Weinheim: Beltz.

IFRC – International Federation of Red Cross and Red Crescent Societies. 2014. *IFRC Framework for Community Resilience.* Geneva, Switzerland: IFRC. https://media.ifrc.org/ifrc/wp-content/uploads/sites/5/2018/03/IFRC-Framework-for-Community-Resilience-EN-LR.pdf

IOHA – International Occupational Hygiene Association. n.d. *What is Occupational Hygiene?* https://www.ioha.net/about/occupational-hygiene/ (accessed May 19, 2020).

Lewis-Smith, H. and K. Sherman. 2019. Body image and appearance-altering conditions. In M. Gellman (Ed.), *Encyclopedia of Behavioral Medicine.* New York: Springer. https://doi.org/10.1007/978-1-4614-6439-6_102007-1. Available from https://uwe-repository.worktribe.com/output/4670426.

Ministry of Education of Portugal. 2017. *Students Profile by the End of Compulsory Schooling.* https://cidadania.dge.mec.pt/sites/default/files/pdfs/students-profile.pdf

OECD - Organisation for Economic Co-operation and Development. 2018. *PISA 2021 Integrated Design*, presented at 46th meeting of the PISA Governing Board (PGB). Available at https://www.oecd.org/pisa/pisaproducts/PISA2021_IntegratedDesign.pdf.

Paulus, P. 2009. *Anschub.de – ein Programm zur Förderung der guten gesunden Schule.* Münster: Waxmann.

Paulus, P and H. Hundeloh. 2020. School heads as change agents: Salutogenic management for better schools. In U. Bollmann and G. Boustras (Eds.), *Safety and health competence. A guide for cultures of prevention*, 199–215. Boca Raton: CRC Press, Taylor & Francis Group.

Presidency of the Council of Ministers. 2018. *Decree-Law no 55/2018*. In Republic Diary No. 129/2018, Series I of 2018-07-06. Available at https://dre.pt/web/en/home/-/contents/115652962/details/normal (accessed May 19, 2020).

Rantanen, J. 2018. Multiple case study on regional OSH networks. In *International newsletter on occupational health and safety, special issue on networking*, 22–25. Geneva: International Labour Organization.

Schleicher, A. 2020. *OECD-Report: Pisa-Studie soll angesichts Corona-Krise soziale Kompetenzen berücksichtigen. Focus.* Available at https://www.focus.de/familie/schule/oecd-report-pisa-studie-soll-angesichts-corona-krise-soziale-kompetenzen-beruecksichtigen_id_11994084.html (accessed May 19, 2020).

Tregenza, T. 2014. *Symposium 5: Networking as a driving force for prevention.* XX World Congress for Safety and Health at Work. Frankfurt: German Social Accident Insurance.

Van Dijk, F. 2017. *Research needs and opportunities for education and training in occupational safety and health.* Speech at ICOH-SCETOH Symposium, Zagreb, October 26, 2017.

13 International Occupational Health Outreach

One Step at a Time – Vietnam Progress Report

Tuan Nguyen
State Compensation Insurance Fund

Mary O'Reilly
University at Albany School of Public Health:

CONTENTS

13.1   INTRODUCTION

Mr. Tuan Nguyen made his first trip back to Vietnam in the December of 1997 after 22 years of living in the United States. He took the opportunity to participate in a study tour to Vietnam, organized by the World Federation of Public Health Associations. His group of 22 public health experts from the United States, Australia, and Thailand had several meetings with Vietnamese professionals in four cities: Ha Noi, Da Nang, Hue, and Ho Chi Minh City. At each stop, they were able to meet with local public health officials and visit health institutions and facilities.

Several meetings were held in Ha Noi (Northern Vietnam) with the Ministry of Health, Ha Noi School of Public Health, and the City Health Department. They also toured the National Institute of Hygiene and Epidemiology (Pasteur Institute), National Institute of Nutrition, Dong Anh District Hospital, Bach Mai Hospital, and the famous Co Loa old rampart. In Da Nang and Hue (Central Vietnam), meetings and site visits were held with local health officials, Hue Medical faculty, Hue Hospital, and Huong Thuy Commune Health Center. In Ho Chi Minh City (Southern Vietnam), several full days were spent in meetings and visits with officials from public health, including the Institute of Hygiene, Public health, Traditional Medicine Faculty, Dentistry Faculty, and Nursing Faculty. Toward the end of the trip, a walk-through evaluation of a lacquerware factory was performed, and several small shops were visited along the way.

Hospitals and clinics, local and international NGOs, and other health facilities operating at the central, district, and commune levels were visited throughout this study tour. The main goals were to establish linkage with professional counterparts in Vietnam and to develop future collaborative activities in mutual learning, scientific exchange, and capacity building.

The trip was eye-opening. Workers in various industries in Vietnam performed their jobs with little or no personal protective equipment. Some lacked the basic knowledge of the health hazards of the materials they worked with. No engineering controls for hazardous processes were seen anywhere. For example, workers making lacquerware sat on solid wood benches while grinding/polishing wooden boxes on stones. Their hand motions were fast and repetitive. In addition, they had to raise their feet to the same height as the benches on which they sat because they had no footrests. This is a very awkward posture. When Tuan tried to mimic their postures and actions on another wooden bench he found that his tailbone tingled after only a few minutes.

At another site, residents were hired to work on refurbishing small lead-acid batteries. Four or five people sat in front of their homes working on batteries. A lady carrying her infant in her arm, and a dog sniffing and walking around, were within a few feet of the batteries.

During the tour, factory owners, the workers, and the public health officials all expressed the desire to learn from the group. They wanted more knowledge in dealing with preventable accidents and injuries. They wanted to learn more about new accident prevention and research techniques, training methods, and the availability of engineering controls and personal protective equipment.

Tuan resolved to return, and hopefully bring whatever knowledge would reduce their risk of disease, and maybe even save a life or two. Occasionally, he explained to the group why the Vietnamese people were reacting in a certain way to foreigners' comments or questions and how to eat certain foods; for that, the group bestowed on Tuan the title "Cultural Ambassador" at the end of the tour with a handmade certificate. As Vietnam continues to modernize, issues such as injury control, occupational health, smoking cessation campaigns, noncommunicable-disease reduction, obesity, and chronic-disease monitoring are very critical to maintaining a healthy workforce.

13.2 OVERVIEW OF THE VIETNAMESE OCCUPATIONAL HEALTH SYSTEM

Currently, occupational safety and health regulations in Vietnam are mainly developed and promulgated by the Ministry of Labor, Invalids and Social Affairs (MOLISA), the Ministry of Health (MOH), and the Vietnam General Confederation of Labor (VGCL), based on authority from Decree 75/2017/ND CP (2017). The MOLISA's responsibilities include management of matters related to the national workforce's state of affairs such as labor code implementation, labor conditions, employment laws, social insurance, safety and labor law training, occupational safety consultation and enforcement, accident statistics reporting, gender discrimination issues, and policies on war veterans. According to Decree 14/2017/ND-CP (2017), the MOH's responsibilities include managing national preventive medicine programs, medical care and treatment, rehabilitation, medical assessment, forensic mental health, psychiatric health, traditional medicine, reproductive health, medical equipment; cosmeceuticals; food safety, health insurance; population; and public health services within the Ministry's jurisdiction. Several areas including training, research, and labor hygiene standards overlap both ministries.

The MOLISA and MOH shoulder the majority of tasks related to the development of laws and regulations governing health and safety in the workplace. These ministries are also responsible for collecting, analyzing, and reporting injury and illness data, developing a depository of health records, and collecting accident statistics. The intricate relationship among the many functions of these two ministries has been deliberately created to foster and ensure cooperation between the two ministries and eliminate rivalry.

As a counterbalance to the MOLISA and MOH, the VGCL is tasked to work closely with employers and trade unions as well as represent workers in salary and working condition negotiations. The goal of the VGCL is to promote a safety culture in the workplace and reduce worker accidents and illnesses (Law on Occupational Safety and Health, 2015).

Under these three national entities of the MOLISA, MOH, and VGCL, there are the Bureau for Safe Work, the National Institute of Occupational and Environmental Health (NIOEH), and the Vietnam National Institute of Occupational Safety and Health, formerly known as National Institute of Labor Protection, respectively. Their functions are to carry out the state mandates of the ministries and trade unions.

The NIOEH was created under the MOH to manage three areas: occupational, environmental, and school health. NIOEH's main functions consist of:

1. Conducting scientific researches on biological and physical hazards in the working environment, environmental health, and school health.
2. Conducting refresher and specialized training for graduate and postgraduate degrees on occupational, environmental, and school health.
3. Coordinating, organizing, and leading the implementation of national guidelines on occupational, environmental, and school health throughout the network.
4. Providing health communication and education to promote community wellness.
5. Collaborating with international organizations to gain and share technical knowledge.
6. Providing technical and consultative services on occupational exposures to industries.

At the national level, the NIOEH is under the Ministry of Health. Three other entities, the Health Environment Management Agency, the Dept. of Occupational Health and Injuries Prevention, and the National Institute of Medical Expertise, as well as three research institutes, are under the MOH. The Science of Occupational Safety and Health Institute (SOSHI) in Ho Chi Minh City is under the Ministry of Labor.

At the provincial level, there are 55 Preventive Medicine Centers with staff specialized in occupational health and about 500 district Centers of Preventive Medicine and District Health Centers. At the Commune level, based on 2016 data, there were a little over 11,000 Health Stations and Medical Units at enterprises (WHO, 2016) (Table 13.1).

13.3 PERSONAL DEVELOPMENT AND INTERNATIONAL COOPERATIVE AGREEMENTS

In October 2006, Tuan attended a wonderful 2.5-day training at the Future Leader Institute sponsored by AIHA. In just a few days, he learned about himself and other volunteer opportunities, as well as how to network and interact with different personalities, and engage with various work styles of peers. He also learned about the best practices of effective managers and leaders. Most importantly, he learned how to build an effective network of mentors and peers. During the training, he visited the McDonald's Hamburger University in Chicago, Illinois, to observe how McDonald's employees and franchisees are trained in various aspects of restaurant management, using a cookie-cutter approach. Their motto, "Learning Today, Leading Tomorrow," really resonated with Tuan.

In 2012, Tuan was appointed as the AIHA Ambassador to Vietnam. This was an opportunity for him to promote the field of industrial hygiene and safety in Vietnam and to encourage public health professionals in Vietnam to join AIHA. He also wanted to jointly organize seminars, workshops, and conferences with various entities in Vietnam to increase the awareness and knowledge sharing in the occupational hygiene field.

After becoming the AIHA Ambassador to Vietnam, he had opportunities to arrange some pro bono training services to Vietnam by volunteer professionals from the United States. In 2015, he coordinated the signing of a memorandum of understanding (MOU) between the AIHA and NIOEH. This MOU signified the willingness of both sides to share ideas, technical knowledge in occupational health research, and injury prevention technologies. With additional assistance, NIOEH would be able to strengthen the occupational and environmental health capabilities in research and training and have access to discounted online educational materials.

In the May of 2016, after many months of working to finalize the Memorandum of Collaboration (MOC), Tuan received an invitation from Dr. John Howard, Director of NIOSH and Dr. Margaret Kitt, Deputy Director of NIOSH, to come to Washington, DC, to attend the signing ceremony of the MOC between the two institutes, at the Centers for Disease Control and Prevention (CDC/NIOSH). The purpose of this MOC is to heighten the cooperation between the two countries in the area of OSH and to improve expertise by sharing knowledge and skills to protect workers and promote best practices to improve worker safety and health.

TABLE 13.1
Summary of Responsibilities and Collaboration

Responsibilities and Collaboration between MOLISA, MOH, VGCL, other ministries, and Trade Unions

- Develop policies and enact legislation relating to the rights and obligations of workers.
- Develop and promulgate national Occupational Safety and Health (OSH) standards.
- Collaborate to inspect, monitor, and supervise the implementation of OSH policies.
- Develop guidance and regulation to improve working conditions.
- Ensure implementation of corrective measures.
- Encourage and provide rewards for workers to follow the rules.
- Work with employers to rescue victims of workplace accidents.
- Promote public engagement to further safe working conditions.
- Identify jobs subject to strict OSH requirements.
- Investigate serious complex occupational accidents.
- Encourage local and international cooperation on OSH.

Responsibilities and Collaboration between MOLISA and MOH

- Investigate accidents, coordinate rescue, and ensure employer fulfills responsibilities.
- Engage the public in workplace safety culture.
- Promulgate a list of jobs subject to OSH requirements.
- Inventory accident reports biannually and annually.
- Identify and inventory all occupational diseases and send them to the Prime Minister annually.
- Set worker compensation for occupational injury and death, and employer obligation to pay.
- Prevent fraud in the worker compensation system.
- Prepare OSH reports on accidents, injuries, and death.
- Manage health effects of foods, pharmaceuticals, vaccines, medical products, cosmetics, drugs, household chemicals, pesticides, and disinfectants.

Responsibilities MOLISA

- Promulgate which supplies and equipment meet OSH standards and how.
- Classify workers by toxic, hazardous, and extremely hazardous working conditions.
- Stipulate regulations on personal protective equipment and hazardous machinery.
- Develop a database on OSH nationwide statistics.
- Develop occupational protections for domestic workers.
- Communicate aspects of OSH legislation and OSH workplace statistics.
- Verify data on accident and illness records.
- Organize and conduct relevant research.

Responsibilities MOH

- Supervise the safety of industrial products such as pharmaceuticals, food, cosmetics, medicines, household chemicals, and pesticides among other things.
- Formulate medical surveillance and return-to-work criteria.
- Evaluate criteria to identify toxic and extremely toxic occupations.
- Collect data on worker-related hospitalizations.
- Organize and communicate prevention of OSH accidents, injuries, and illnesses.
- Develop appropriate OSH training.

Source: Law on Occupational Safety and Health (2015).

It was a wonderful occasion, with presentations in person from Dr. Margaret Kitt, Dr. Nguyen Thai Hiep Nhi, Vice Head of Planning and International Collaboration, Dr. Leslie Nickels, and via videoconferencing from Dr. Marie Sweeney, Susan Moore, Jessica Ramsey, and Max Kiefer. In between these presentations, the signing of the MOC was finalized.

Witnessing the signing of the MOC by Dr. John Howard, Director of NIOSH, and Dr. Doan Ngoc Hai, Director General of NIOEH, was a memorable experience for all the delegates from both Vietnam and the United States. Much to everyone's surprise and delight, a ceramic tile montage on the wall in the courtyard of the institute with pictures of doctors Doan Ngoc Hai and John Howard to commemorate the signing ceremony greeted the group teaching in Ha Noi in 2018.

In the March of 2018, Dr. Margaret Kitt, Deputy Director of NIOSH, invited Tuan to come to Morgantown, West Virginia, to meet with members of the Ho Chi Minh City SOSHI. Attendees at that meeting had the privilege to hear from various staff members about NIOSH functions, specialties, and emphasis on occupational health areas. Dr. Kitt gave an overview of the NIOSH's functions and expertise. Other presentations were made by Lisa Delaney, Associate Director of the Emergency Preparedness and Response Program, and Dawn Castillo, Director of the Division of Safety Research, explaining the multidiscipline focus on basic, applied, and preventive laboratory research for controlling and preventing workplace safety and health problems. Meetings were held with John Noti, Chief of Allergy Clinical Immunology Branch, Mike Andrew, Chief of Biostatistics & Epidemiology Branch, Robert Lanciotti, Acting Chief of Exposure Assessment Branch, Ren Dong, Chief of Engineering and Control Technology Branch, and Jeff Fedan, Chief of Pathology and Physiology Research Branch. In the presentation, the Director of SOSHI expressed the desire to share and exchange OSH-related information and resources and to collaborate on scientific activities in which both Institutes have a common interest, such as training courses or seminars.

A few months later, the MOC signing ceremony between the NIOSH and SOSHI was held in Ho Chi Minh City. Within 2 years, both the NIOEH in Ha Noi and the SOSHI in Ho Chi Minh City had MOCs with NIOSH/CDC in the United States.

Immediately following the signing of the MOC between SOSHI and NIOSH, an international workshop was held in a newly constructed building at SOSHI in the September of 2018. Dr. Kitt and Dr. Lincoln talked about the future of OSH, and Tuan talked about the roles of the federal and state government in managing and guiding safety culture development. Participants at the workshop included leaders of state management agencies in charge of occupational safety and hygiene and domestic scientists in the field of occupational safety of Vietnam. The presence of NIOSH scientists and other experts in the occupational safety field from the United States, Korea, Malaysia, and Australia showed that occupational safety in Vietnam is garnering attention globally. The need for building a safety culture in all workplaces to improve working conditions, prevent labor accidents, and reduce occupational diseases was highlighted and emphasized. The workshop demonstrated that the success of a workplace culture of safety contributes to an increase in productivity and an enhancement of the business's reputation and brand. The events were televised and published in local newspapers.

13.4 FOSTERING INTERNATIONAL SCIENTIFIC CONFERENCES

Over the years, Mr. Tuan Nguyen held some informal technical exchanges with several staff members of the NIOEH and other universities. In 2005, he attended the Second Scientific Conference on Occupational and Environmental Health hosted by the NIOEH (Ministry of Health) in Ha Noi. At this conference, he shared the workers' compensation insurance costs for occupational injuries from State Compensation Insurance Fund's claims data for the electroplating industry. The study examined the frequency distribution of different types of claims and their respective costs for the electroplating classification. The causative factors for these types of claims were identified and evaluated for appropriate intervention approach planning.

During the same year, he attended a Water pollution prevention technologies conference hosted by the Vietnam Education Foundation and the Vietnamese Academy of Science and Technology through an invitation from a friend. Several informal visits to Hue University in Central Vietnam were made during Tuan's trips to China in 2011 and 2013.

At the end of June 2016, after completing a week of delivering training at the NIOEH, Tuan attended the First Asian Network of Occupational Hygiene (ANOH) in Ha Noi. There were more than 100 delegates from various parts of the world in attendance. At this conference, Dr. Doan Ngoc Hai and Noel Strider received awards for their contributions in the field, and a member of the newly formed Vietnamese Industrial Hygiene Association became a member of the governing board of ANOH.

The 2017 American Industrial Hygiene Conference and Exposition (AIHce), an international conference, was held in Seattle. Abstracts from the US branch of Workplace Health Without Borders (WHWB) were accepted on Vietnam's unique approach to exposures associated with industrial operations performed by specialized villages. The technical session included Dr. Doan Ngoc Hai, Mr. Tuan Nguyen, and Dr. Mary O'Reilly. Their presentation is summarized in the following case.

Case:

The village of Dong Mai, Hung Yen province, has been recycling batteries since the 1980s. Recycling activities were done in residential areas and caused environmental pollution on a large scale, especially soil pollution. Residents, including children, have elevated blood lead levels. Contaminated soil particles became airborne when disturbed, further increasing exposure.

NIOEH collaborated with a number of domestic and international agencies and organizations to conduct research and intervention studies on the Dong Mai lead poisoning issues. Some interventions included remediation of contaminated soil, relocation of production establishments out of the residential areas, and relocated residents to another area.

Despite their best efforts, all the lead could not be removed or remediated in place as desired, and the inhabitants could not be relocated, as is often done in the United States. The Vietnamese have chosen instead to explore intervention techniques not embraced by the developed world in the occupational and environmental health community. Although exposure to lead is a global problem, solutions work best when they are compatible with local customs, resources, and traditions. Sometimes the best solution is a combination of approaches used simultaneously and/or sequentially, especially when resources are limited.

After moving the process of dismantling acid-lead batteries to an industrial zone, children and exposed workers were treated with Pectin Complex products. Pectin works by binding metals in the digestive tract to produce insoluble compounds that are not absorbed. The lead bound to the pectin is excreted in the body waste excrement. (Niculescu et al, 1969; Rowland et al, 1986). Although lead poisoning was not completely remediated, 65.25% children had decreased blood lead levels (BLL), 24.58% had unchanged BLLs and 10.17% had increased BLLs. Some of the increased BLLs in children were due to a continuing low-level exposure and/or a failure to follow recommended dosing with pectin. Over 95% of workers had decreased BLLs, 2.3% increased slightly, and 2.3% had no change. (Hai and Tung, 2017)

Using Pectin Complex according to the guidelines in combination with enhanced health education to prevent lead poisoning has brought significant decreases in blood lead levels, despite the high number of children and workers with BLLs above permissible limits which require further intervention. NIOEH continually conducts periodic health check-ups and blood lead testing for children and employees to study the effectiveness of Pectin products for lead poisoning prevention.

Immediately after the AIHce in Seattle in 2017, Tuan was able to arrange a meeting with both the Center for Occupational and Environmental Health and the University of California Irvine Division of Occupational and Environmental Medicine. It was a warm and fruitful meeting for both Dr. Doan Ngoc Hai and Tuan. They were able to tour the Center of Occupational and Environmental Health Clinic in the morning; then, in the afternoon, we attended a resident teaching session called Journal Club. They also visited the Air Pollution Health Effects Laboratory of the School of Medicine.

The Fifth International Scientific Conference on Occupational Health, in which NIOSH/CDC was one of the main event sponsors, was held in September 2018 in Ha Noi, Vietnam. Dr. Doan Ngoc Hai was the chair, and Tuan served as the International Liaison for the conference. The organizing committee team consisted of Dr. Nguyen Bich Diep, Dr. Doan Ngoc Hai, and Dr. Nguyen Van Son of NIOEH from Vietnam, and Dr. Margaret Kitt, Tricia Boyles, and Debbie Hoyer from the United States. It was an exciting and extremely busy time for everyone, looking for conference sponsors, exhibitors, and volunteers to be a part of this international conference. The conference committee also developed a 3-day conference program, inviting keynote speakers, session chairs, and guest speakers from around the world to take part in this conference. They reviewed over 100 abstracts, translated them into both languages – Vietnamese and English – and were delighted to have NIOSH/CDC, WHO Western Pacific Region, SKC Asia of Singapore, and Rion of Japan as sponsors of the event.

This conference examined workplace safety and environmental health conditions and described methods to evaluate and control hazards. The conference's theme was "Occupational Health and Environment: Challenges and Opportunities in Sustainable Development." This international conference was an opportunity for local and international scientists to exchange information and experiences on research studies in occupational and environmental health, school health, and intervention measures on health promotion and protection for workers, school children, and community. The host of this event was the NIOEH, a unit under the Vietnam Ministry of Health and a WHO (World Health Organization) Collaborating Center for Occupational Health.

Finally, after many months of hard work of the organizing committee, the event started without a glitch. It was well attended with more than 300 national and international scientists working in the fields of OSH, Environmental Health, School Health, Public Health, etc. After 2.5 days of convening, it was announced as a successful event by creating an atmosphere of solidarity, openness, and responsible science. All of the conference abstracts were published in the *Journal of Health and Pollution* (2019), a journal funded by the World Bank and the European Union.

13.4.1 TRAINING ACTIVITIES

While attending the AIHce Conference in Salt Lake City, Utah, in 2015, Tuan was introduced to Marianne Levitsky, President of WHWB, by a great colleague, Mark Katchen, President of the Phylmar Group. Tuan told Mark about what I was doing in Vietnam and he suggested working with Marianne's group since her nonprofit group does work internationally. Tuan also talked to Roger Alesbury, and other members of the British Occupational Health Society and Occupational Health Training Association (OHTA) based in the United Kingdom, regarding the use of online training courses to develop an internationally recognized certification program for Vietnam Occupational Health society professionals. By being an AIHA Ambassador to Vietnam and a member of WHWB, Tuan received many opportunities to meet with health and safety professionals from across the country and around the world to share insights, tackle challenges, and partner in the pursuit of the best industrial hygiene and safety standards.

Shortly after coming back from signing the MOC between NIOSH and NIOEH in 2016, an official invitation to come to Vietnam to conduct training from June 20 to 24, 2016 was offered. Although planning this activity for several months, there were about 2 weeks to finalize all the logistics. Marianne was extremely helpful in helping to invite additional trainers, including Dr. Elain Lindars and Noel Tresider, OHTA Board member from Australia, and Jonathan Haney and Dr. Mary O'Reilly from the United States. All of the trainers, including Tuan, were members of WHWB. Since this was the first training in Vietnam, no one knew how many people would attend the training even though the institute had been asked to arrange for about 25–30 attendees.

Amazingly, all the trainers showed up as scheduled in Ha Noi. Although they were communicating through emails for a few months, everyone was so happy to finally meet in person. The staff at NIOEH were so punctual and hospitable. They picked everyone up at the airport and brought them to their hotels. After a few hours of rest, everyone met with other staff members of the institute for a quick meeting and also went to the local phone shops to get SIM cards to communicate with each other. The group planned to be in Ha Noi 2 days before the training to finalize all the details before the class started.

The institute decided to invite a team of industrial hygienists and scientists to help them further develop the occupational hygiene capacity in the country. With the support of Dr. Hai and Dr. Hiep Nhi, we were able to organize the course to train the staff of the institute at their premises in Ha Noi, Vietnam.

The OHTA W201 training module was selected for this 5-day course. It was taught in English. This course was the first to be taught in Vietnam and, therefore,

FIGURE 13.1 Opening ceremony in 2016 Dr. Hai addressing the class. The instructors: Jonathan Haney, Dr. Mary O'Reilly, Tuan Nguyen, Dr. Elaine Lindars, Noel Tresider. (Photo courtesy of Tuan Nguyen.)

has provided a stepping stone for presenting future OHTA courses to help fill the knowledge gap as Vietnam increases industrial production. A few modifications were made to the course to enhance the interest and knowledge level of the attendees. The added changes included a site visit to a fan factory and a section on industrial hygiene problems in the oil and gas industry. From time to time, Tuan translated and explained some terminologies used in the presentation or discussion to ensure that we all were on the same page. In Figure 13.1 above, Dr. Hai Ngoc Doan introduced the panel of instructors during the opening ceremony of the OHTA W201 class.

A full class was expected to be about 30 students but there were over 50 attendees when the class was actually given. The additional attendees included a representative from the Cambodian Department of Safety & Health, Ministry of Labor & Vocational Training, and a Laotian Technical Officer in OSH, Ministry of Labor and Social Welfare. Other attendees were from the Ha Noi Medical University, several companies, and various other government agencies. The majority of the attendees had degrees in either medicine or public health fields,, such as MDs, PhDs, MPHs, and MSc. All could read and write English; however, English oral skills varied from poor to excellent.

A site visit tour was hurriedly planned for the students to provide experience in a walk-through survey of a factory that- manufactures small domestic fans. After about half an hour, however, the visit was curtailed by the factory management since the group was too big and had asked too many questions.

After the visit to the factory, Tuan and Noel Tresider led the discussion with the students to get their observations and comments. Although the visit was short, the students split into about ten different groups with each group pursuing a different aspect of the factory work. Collectively, the students were able to obtain interview and observation data equivalent to a full-day survey. It was a lively discussion with much interaction as the class debated and shared the observed working conditions, hazards, as well as existing and potential control measures.

At the conclusion of the course, all participants were presented with a certificate by the trainers, and Tuan was able to invite the Health Affairs Attaché of the US Embassy in Vietnam, Dr. Jeffrey O'Dell, to attend and participated in the ceremony.

As a follow-up to this initial occupational health training in 2016, two courses were offered in the September of 2018. From Thursday, September 13, through Tuesday, September 18, eight volunteers from WHWB-US and WHWB worked with 54 people in Ha Noi and 35 people in Ho Chi Minh City to improve their knowledge about measuring hazardous substances in the workplace.

The course in Ha Noi followed the syllabus of the OHTA for the W501 course in the Measurement of Hazardous Substances, complete with case studies and practical exercises. The equipment was provided by SKC Asia. Many of the students in the Ha Noi class had advanced degrees such as masters, MDs, and PhDs, so they were well versed in the theoretical and scientific aspects of occupational hygiene. They were interested in, and benefited from, the practical, hands-on experience they got during the class exercises. Figure 13.2 shows a lunch meeting among all the instructors and guests, a day before the start of the training in Ha Noi.

The course in Ho Chi Minh City was a 3-day course (Friday–Sunday, September 14–16) on theoretical (risk assessment, biological monitoring) and practical (exposure monitoring equipment) aspects of occupational hygiene. After the first day of teaching in Ha Noi, three of the instructors, along with Dr. Diep from the NIOEH, traveled to Ho Chi Minh City to start the class there. After teaching on Friday, two instructors flew back to Ha Noi and were replaced by two additional instructors from Ha Noi. At the end of the 3-day class in Ho Chi Minh City, all the instructors flew back to Ha Noi to conclude the 5-day course and help administer the exam. Many of the students in the Ho Chi Minh course already worked in industry; others worked for a variety of government ministries and institutes. Each of the students in both courses received a certificate of completion with WHWB, AIHA, and NIOEH logos.

FIGURE 13.2   The instructors in the 2018 class from right to left are: Ted Zellers, PhD, Wendy Zellers (guest), Warren Silverman, MD, Jonathan Haney, CIH (ret), Albert Tien, PhD, Fred Mlakar, CIH, CSP, Lauralynn McKernan, PhD, CIH, Tuan Nguyen, MBA, CIH, CSP, ARM, FAIHA, Course Coordinator, and Mary O'Reilly, PhD, CIH, CPE, FAIHA, Course Director. (Photo courtesy of Tuan Nguyen.)

13.5   CONCLUSIONS

In order to share the research knowledge and technologies in OSH with Vietnam, a strong foundation of knowledge has to be built by reaching out to all experienced researchers, professors, professionals, as well as universities, governmental institutions, and NGOs. An all-inclusive approach is needed to be put in place to ensure that the efforts would pay off and be sustainable.

The quest into international engagement between various bodies of knowledge in the OSH field is an interesting and productive journey, with many unexpected outcomes and rewards. Along the way, many benefactors have been met and many friends made. Everyone was touched by the generosity, commitment, and kindliness of all the volunteer trainers in their contribution to this endeavor. All the associations, collaborators, and institutes that support the increasing OSH awareness in this part of the world are essential for success.

The goals to share and transfer OSH knowledge and technologies with developing countries to ensure that the workers can have a safe workplace and prevent accidents align with the vision and mission of AIHA and WHWB. Although the Vietnamese public health professionals want to learn from western industries, the industrial world can also learn from them regarding their efforts and achievements with very limited resources. In the end, the international workforce needs to know that occupational health advocates care and that communication and commitment can bring about increased workplace health.

REFERENCES

Hai, D. N., Tung, L. V. 2017. The effectiveness of intervention for lead poisoning prevention in workers and children at a battery recycling village in Hung Yen province in 2016. Technical Session Presentation at AIHCe.

National Assembly Socialist Republic of Vietnam. 2015. Law on Occupational Safety and Health, No. 84/2015/QH13.

Niculescu, T., Rafaila, E., Eremia, R., Balasa, E. 1969. Investigations on the action of pectin in experimental lead poisonings. *Rom Med Rev* 13: 3–8.

Occupational Health and Environment: Challenges and Opportunities in Sustainable Development. 2019. *J Health Pollut* 9(23): 190901. https://www.journalhealthpollution.org/doi/pdf/10.5696/2156-9614-9.23.190901 (accessed June 28, 2020).

Rowland, I., Mallett, A., Flynn, J., Hargreaves, R. 1986. The effect of various dietary fibres on tissue concentration and chemical form of mercury after methylmercury exposure in mice. *Arch Toxicol* 59: 94–98.

Socialist Republic of Vietnam. 2017. Duties and responsibilities (Decree 14/2017/ND-CP. 2017). http://english.molisa.gov.vn/Pages/About/DutiesResponsibilities.aspx (accessed June 28, 2020).

WHO. 2016. *Human resources for health country profiles: Viet Nam, World Health Organization Regional Office for the Western Pacific*, World Health Organization, ISBN 978 92 9061 771 6 (NLM Classification: W76 JV6)

14 Pathways Forward in New Collaborations for Capacity Building in Occupational Hygiene

Thomas P. Fuller
Illinois State University and International
Occupational Hygiene Association

CONTENTS

14.1 INTRODUCTION

The final contract for this book was approved in the November of 2019, with subsequent edits based on reviewer comments and changes to the Table of Contents and author list being completed in early 2020. Authors were assigned, and about half met their due date of May 1 for their respective chapters. As expected, the remainder trickled in over the next few months. The conclusion was written in the final months of November and December 2020 and, therefore, had the benefit of knowing what had happened over the past 9 months with regards to the global pandemic of COVID-19.

COVID-19 has changed much in society and has altered how we think of the world and occupational health. Collaborations and networking described in the prior chapters of this book were overturned throughout the year, with conferences canceled, travel for research and training projects curtailed, and many normal associations and collaborative projects decreased in scope.

However, most groups learned to adapt quickly. Conferences went virtual, and meetings were held via internet hosting platforms. In some cases, such as the annual IOHA board meeting, attendance and participation in the virtual meeting was actually higher than in previous years, since the expense of global travel and lodging was eliminated. Groups who previously conducted monthly board meetings became proficient in hosting virtual meetings on a regular basis, and if anything, connectivity between members increased.

Rather than traveling to foreign countries to perform training, e-learning platforms were expanded as a means to provide courses, normally taught by instructors physically, to be taught online. Although there are indications that the effects of the pandemic will decrease in future months with the distribution of vaccines, it seems that the use of e-learning tools and platforms will continue to increase, even after travel is possible again.

14.1.1 Contributions of Occupational Hygiene

Early on in the pandemic, occupational hygienists stepped up to support workers on the front lines. Occupational hygienist expertise in ventilation, respiratory protection, personal protective equipment, and communicable disease transmission was mobilized by individuals and groups. The American Industrial Hygiene Association (AIHA) created an ad hoc COVID Response Team comprised of members from various AIHA committees such as the Respiratory Protection Committee, the Health Care Working Group, the Nonionizing Radiation Committee, and the Biological Safety Committee. This group met via video conferencing regularly and worked to create a variety of fact sheets and other public information materials based on occupational hygiene principles of practice. AIHA then distributed these documents publicly via various outlets. Simultaneously, AIHA formed another group of members to create a series of "Back to Work Safely" guidelines meant to be used by businesses and consumers to safely reopen from COVID-19 quarantines (AIHA, 2020).

In September 2020, the total health care worker deaths due to COVID-19 exposure in the workplace globally were estimated to be over 7,000 and rising. A breakdown of some of the nations contributing to this number is shown in Table 14.1 (Amnesty International, 2020). But, in addition, frontline workers in education, service industries, food production, nursing homes, law enforcement, and corrections are all being

TABLE 14.1
COVID Deaths in Health care Workers

Country	Deaths
Mexico	1,320
United States	1,077
United Kingdom	649
Brazil	634
Russia	631

exposed and infected with the virus. Many, if not most, of the illnesses that result from the exposures, illnesses, hospitalizations, and deaths are not being counted as workplace related.

Throughout the pandemic, occupational hygienists have contributed to an understanding of the transmission of the SARS-CoV-2 virus in the environment. Numerous publications have been forthcoming regarding the ability of the virus to remain viable in the air and travel several feet for many hours to infect others in the general vicinity (Bahl et al., 2020; Almilaji and Thomas, 2020; Covaci, 2020). Webinars on infectious disease transmission, methods of disinfection and control, and the use of PPE and respiratory protection have been taught by occupational hygienists. As we move forward through the pandemic, occupational hygienists will continue to research and shed light on better means of protection and control. They are also working on how to handle and respond better to the next inevitable infectious disease outbreaks.

In addition to responding to COVID-19 as an infectious agent and workplace hazard, occupational hygienists should be preparing themselves as a profession for the coming onslaught of post-COVID workers who will be disabled by the disease in a broad range of ways and seriousness. There are already known respiratory consequences for large numbers of workers who may need special compensations to do work that had been done previously, or possibly reassigned new jobs altogether. Other disabling injuries may include neurological disorders and heart and circulatory problems, psychological disturbances, organ damage, and joint pain. Occupational hygienists will need to play a greater role in the evaluation and designation of worker activities post-COVID.

14.2   OTHER CAPACITY BUILDING COLLABORATIVE ORGANIZATIONS AND PROJECTS MOVING FORWARD

Several other collaborative projects not discussed in earlier chapters, or only just being planned, are briefly summarized in the next section. The bottom line is that occupational-hygiene collaboration will continue, however the format. We have learned how to maintain and even expand communications and collaborations, even amid a global pandemic that curtails physical travel.

14.2.1   WORLD HEALTH ORGANIZATION COLLABORATING CENTERS FOR OCCUPATIONAL HEALTH

In 2015, the WHO celebrated the 25th anniversary of the Network of Collaborating Centers for Occupational Health and the mission of stimulating associations with international partners in improving occupational safety and health. At that time, the network included 55 collaborating centers, including the International Commission on Occupational Health, the International Occupational Hygiene Association, the International Ergonomics Association, and the International Labour Organization (ILO). The network was also supported by the National Institute of Occupational Safety and Health, USA, and the Finnish Institute of Occupational Health (WHO, 2016).

Over the 25-year period from 1990 to 2015, the network developed and implemented a comprehensive plan of action to cover the following topics:

1. To devise and implement policy instruments on workers' health.
2. To protect and promote health at the workplace.
3. To improve the performance of and access to occupational health services.
4. To provide and communicate evidence for action and practice.
5. To incorporate workers' health into other policies and projects.

In addition to the overall subject topics, specific objectives and development priorities were created in the following areas (NIOSH FIOH, 2016).

Global Plan of Action Objective 1: To Devise and Implement Policy Instruments on Workers' Health

Priority 1.1: Develop/update national profiles on workers' health and national action plans on workers' health.

Priority 1.2: Develop and disseminate evidence-based prevention tools and raise awareness for the prevention of dust-related diseases.

Priority 1.3: Develop and disseminate evidence-based tools and raise awareness for the elimination of asbestos-related diseases.

Priority 1.4: Conduct studies and develop evidence-based tools and information materials for the comprehensive protection and promotion of health for health care workers.

Global Plan of Action Objective 2: To Protect and Promote Health at the Workplace

Priority 2.1: Practical toolkits for assessment and management of occupational health risks.

Priority 2.2: Healthy Workplace programs and guidance.

Priority 2.3: Toolkits for the assessment and management of global health threats (including HIV, tuberculosis, malaria, influenza).

Global Plan of Action Objective 3: To Improve the Performance of and Access to Occupational Health Services

Priority 3.1: Develop working methods and provide technical assistance to countries for the organization, delivery, and evaluation of Basic Occupational Health Services in the context of primary health care.

Priority 3.2: Adapt and disseminate curricula, training materials, and training for international capacity building in occupational health.

Global Plan of Action Objective 4: To Provide and Communicate Evidence for Action and Practice

Priority 4.1: Encourage practical research on emerging issues.

Priority 4.2: Further develop the global research agenda for workers' health.

Priority 4.3: Revision of the International Statistical Classification of Diseases and Related Health Problems to include occupational causes.

Global Plan of Action Objective 5: To Incorporate Workers' Health into Other Policies and Projects

Priority 5.1: Collate and conduct cost–benefit studies to clarify the economic benefits of workers' health.

Priority 5.2: Develop specific and relevant recommendations to manage risks associated with the impacts of globalization on workers' health.

Priority 5.3: Implement toolkits for the assessment and management of OSH hazards in high-risk industry sectors and vulnerable worker groups.

14.2.2 VISION ZERO

In 2015, the ILO launched the Vision Zero Fund as an initiative of the Group of 7 countries to work together to prevent work-related deaths, injuries, and diseases in sectors operating in, or hoping to join, global supply chains (GSCs). The main objectives were to expand public and private collaborative activities to improve and enhance occupational safety and health (OSH) prevention activities in businesses operating in low- and middle-income countries (ILO, 2016).

The Vision Zero approach adopts a strategy and intervention framework with the goal of eliminating severe or fatal work-related accidents, injuries, and diseases in GSCs. A fund was created that is supported by nations, intergovernmental and nongovernmental organizations, private sources, other foundations, and individuals. Funds are to support various projects operating under the umbrella of "Vision Zero: Achieving a world without fatal or serious occupational accidents and diseases" (ILO, 2019).

The ILO reported that, in the past, many collaborative efforts to expand OSH capacity and improve working conditions for employees around the world tended to be episodic and inconsistent. Many gaps in worker health and safety still remain in several regions of the world. The ILO noted that in order to make substantial and permanent improvements in worker rights to OSH, greater efforts need to be made collectively and involve the action of all major stakeholders in global supply chains.

A multi-stakeholder approach that involves governments, workers and trade unions, employers and their organisations, multilateral organisations, civil society and development **agenci**es, working together so that each meets its responsibilities consistent with organizational roles, to implement an agreed plan or set of actions to reduce severe or fatal work accidents, injuries or diseases in global supply chains.

ILO (2016)

The ILO strategy includes obtaining commitments for financial support and action from global companies and relevant stakeholders. It also involves creating and sharing a transparent knowledge base on OSH between collaborators. Then, based on the consistent knowledge base, programs would put forward action plans and other projects toward achieving goals. Ultimately, the strategy will lead to consensus and agreements about the implementation of OSH in GSCs and a broad range of participant and stakeholder levels.

Interestingly enough, the chapters and activities that have been presented in this book seem to represent what the ILO refers to as "episodic improvements" in OSH supply and capacity. Many of the activities the ILO is promoting in the Vision Zero plan are been being done for the past two decades by many of the groups described in this book; these are:

- Direct interventions at factory/production/enterprise level
- Research and knowledge generation
- Legal and policy support and development
- Knowledge sharing and awareness raising
- Capacity building
- Institutional development

Hopefully, moving forward, these groups will be able to collaborate further and find ways to be involved in Vision Zero projects and funding sources. Hopefully, the members of the organizations described in the book will find ways to collaborate and link activities with the ILO approaches and vision.

14.2.3 Global Coalition on Occupational Safety and Health

In the November of 2019, the ILO and several other governmental, nongovernmental, and tripartite organizations established a coalition intended to collaborate on projects toward the improvement of worker health and safety, globally. The coalition is divided into task groups intent on working on various priorities of the coalition agenda. One task group will work on the development of a multiregional OSH information system and database. The second task group will work on the implementation of the ILO Vision Zero program at the enterprise level. The third task group is charged with finding ways to combat noncommunicable diseases such as cancer and cardiovascular diseases that result from workplace exposures. The fourth task group is charged with the promotion of workplace safety for migrant workers. And the last task force has been requested to promote decent worker and productive employment through higher education (ILO, 2020).

In each task force, the participating organizations will send representatives to collaborate and identify goals, projects, and timelines in which to achieve them. The coalition assumes the notion that the process and the outcomes will expand awareness of these important issues and lead to meaningful and sustainable solutions. In this first phase of the coalition, the plan is to report the successes of each task force at the next Global Coalition summit in Japan in 2022. Sessions are currently being planned and speakers identified for the various sessions. The tentative list of sessions include;

- Prevention Culture of Vision Zero
- Vision Zero in Supply Chains of Emerging Nations
- Vision Zero Implementation in Enterprises
- Management and Leadership
- Education/E-learning/Qualification

- Robotics and Collaborative Safety
- Artificial Intelligence, Information Technology, and Digitalization
- Mobility and Auto-Guided Vehicles
- International Standardization for Safety, Health, and Well-Being
- Occupational Safety and Health and the Zero Accident Campaign in Manufacturing
- Occupational Safety and Health in Construction and Civil Engineering
- Occupational Safety and Health in Agriculture
- Occupational Health and Hygiene and Infectious Diseases
- Sustainable Development

14.2.4 INSTITUTE FOR HEALTH METRICS AND EVALUATION (IHME)

The Institute for Health Metrics and Evaluation (IHME) is an independent global health research center founded in 2007, with support from the Bill and Melinda Gates Foundation at the University of Washington, USA. The goal is to provide an impartial, evidence-based picture of global health trends to inform the work of policymakers, researchers, and funders. The group works with a broad network of researchers, statisticians, and policymakers, following a science-based, collaborative approach. They foster a transparent and constructive dialogue and debate about all aspects of health measurement (IHME, 2020). http://www.healthdata.org/

Collaboration plays a critical role in the IHME study of the global burden of disease. IHME engages a large network of collaborators with specialties in various topic areas including demographics, risk assessment, injury and illness, epidemiology, health policy, and occupational exposure assessment. An online portal to apply as an individual or organizational collaborator can be found on the IHME website.

14.3 CONCLUSION

Hopefully, OSH professionals will read this book and, then, become aware of the many opportunities available for collaboration and capacity building. If the goal of the ILO is to be the hub of the stakeholders for the Vision Zero strategy, perhaps the information here in this book, all in one place, will be useful to them moving forward.

It is often difficult to grasp how international organizations work together and stay up-to-date on all of the activities and associations. This book should be a resource that professionals and students can use to begin to understand the inner workings of some of these groups and identify ways that they may interconnect and become involved.

REFERENCES

AIHA, Back to Work Safely, AIHA, Falls Church, VA (2020) acquired at https://www.back-toworksafely.org/. Accessed December 28, 2020.
Almilaji, O., Thomas, P., Air Recirculation Role in the Infection with COVID-19, Lessons Learned from Diamond Princess Cruise Ship, *medRxiv* (July 9, 2020) acquired at https://www.medrxiv.org/content/10.1101/2020.07.08.20148775v1. Accessed December 29, 2020.

Amnesty International, Exposed, Silenced, Attacked: Failures to Protect Health and Essential Workers during the COVID-19 Pandemic, Amnesty International, London (2020). Index Number: POL 40/2572/2020.

Bahl, P., Doolan, C., de Silva, C., Chughtai, A., Bourouiba, L., MacIntyre, C., Airborne or Droplet Precautions for Health Workers Treating COVID-19?, *Journal of Infectious Diseases* (April 16, 2020) acquired at https://academic.oup.com/jid/advance-article/doi/10. 1093/infdis/jiaa189/5820886?login=true. Accessed July 28, 2020.

Covaci, A., How can airborne transmission of COVID-19 indoors be minimized?, *Environment International* (2020) Vol 142, 105832.

IHME, Institute for Health Metrics and Evaluation (2020) acquired at http://www.healthdata.org/. Accessed December 29, 2020.

ILO, VISION ZERO FUND: Achieving a World without Occupational Accidents and Diseases (2016) acquired at https://www.ilo.org/wcmsp5/groups/public/---ed_dialogue/---lab_admin/documents/projectdocumentation/wcms_572474.pdf. Accessed December 16, 2020.

ILO, Vision Zero Fund, Strategy 2019–2023, Collective Action for Safe and Healthy Supply Chains (2019) acquired at https://www.ilo.org/wcmsp5/groups/public/---ed_dialogue/---lab_admin/documents/publication/wcms_729031.pdf. Accessed December 16, 2020.

ILO, The Global Coalition for Safety and Health at Work: Task Groups (April 7, 2020) acquired at https://www.ilo.org/global/topics/safety-and-health-at-work/programmes-projects/WCMS_740999/lang--en/index.htm. Accessed December 24, 2020.

NIOSH FIOH, Improving Workers' Health Across the Globe – Advancing the Global Plan of Action for Worker's Health, NIOSH US and FIOH (May 2016) DHHS (NIOSH) Publication No. 2016-118.

WHO, Global Network of WHO Collaborating Centres for Occupational Health, WHO World Press, Geneva (2016).

Index

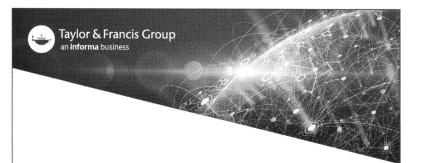

Printed in the United States
by Baker & Taylor Publisher Services